T0141875

Sustainable Civil Infrastructures

Editor-in-Chief

Hany Farouk Shehata, SSIGE, Soil-Interaction Group in Egypt SSIGE, Cairo, Egypt

Advisory Editors

Khalid M. ElZahaby, Housing and Building National Research Center, Giza, Egypt
Dar Hao Chen, Austin, TX, USA

Sustainable Infrastructure impacts our well-being and day-to-day lives. The infrastructures we are building today will shape our lives tomorrow. The complex and diverse nature of the impacts due to weather extremes on transportation and civil infrastructures can be seen in our roadways, bridges, and buildings. Extreme summer temperatures, droughts, flash floods, and rising numbers of freeze-thaw cycles pose challenges for civil infrastructure and can endanger public safety. We constantly hear how civil infrastructures need constant attention, preservation, and upgrading. Such improvements and developments would obviously benefit from our desired book series that provide sustainable engineering materials and designs. The economic impact is huge and much research has been conducted worldwide. The future holds many opportunities, not only for researchers in a given country, but also for the worldwide field engineers who apply and implement these technologies. We believe that no approach can succeed if it does not unite the efforts of various engineering disciplines from all over the world under one umbrella to offer a beacon of modern solutions to the global infrastructure. Experts from the various engineering disciplines around the globe will participate in this series, including: Geotechnical, Geological, Geoscience, Petroleum, Structural, Transportation, Bridge, Infrastructure, Energy, Architectural, Chemical and Materials, and other related Engineering disciplines.

More information about this series at http://www.springer.com/series/15140

Louay Mohammad · Ragaa Abd El-Hakim
Editors

Sustainable Issues in Transportation Engineering

Proceedings of the 3rd GeoMEast
International Congress and Exhibition, Egypt
2019 on Sustainable Civil Infrastructures –
The Official International Congress
of the Soil-Structure Interaction Group
in Egypt (SSIGE)

 Springer

Editors
Louay Mohammad
Louisiana State University
Louisiana, LA, USA

Ragaa Abd El-Hakim
Tanta University
Tanta, Egypt

ISSN 2366-3405   ISSN 2366-3413   (electronic)
Sustainable Civil Infrastructures
ISBN 978-3-030-34186-2   ISBN 978-3-030-34187-9   (eBook)
https://doi.org/10.1007/978-3-030-34187-9

This Springer imprint is published by the registered company Springer Nature Switzerland AG
The registered company address is: Gewerbestrasse 11, 6330 Cham, Switzerland

Contents

About the Editors

Dr. Louay Mohammad is a national and international expert in the area of pavement materials and sustainable asphalt construction. He is the holder of the Irma Louise Rush Stewart Distinguished Professor and the Transportation Faculty Group Coordinator at Louisiana State University (LSU). He also serves as Director of the Engineering Materials Characterization and Research Facility at the Louisiana Transportation Research Center (LTRC). He teaches and conducts research in the areas of highway construction materials, pavement engineering, accelerated pavement testing, advanced material characterization and modeling, and infrastructure sustainability. He has served as PI or Co-PI on more than 58 research projects (NCHRP Project 9-40 and 9-40A, NCHRP Project 9-48, NCHRP 9-49A, NCHRP 10-84, NCHRP Project 20-07/Task 361, etc.) totaling over US $12.4 million. He has authored/co-authored more than 270 publications in pavement engineering including over 150 refereed papers and delivered over 170 keynote and invited presentations at national and international conferences. He has developed many standard test methods (AASHTO TP 114, AASHTO TP 115, and Louisiana DOTD TR 330) and mechanistic models that have impacted pavement material characterization and performance, and contributed to changes of asphalt specifications. He is Chair of ASTM subcommittee D 4.25 on Bituminous Mixture Analysis, Past Chair of the TRB Committees AFK40 on Characteristics of Bituminous-Aggregate Combinations to Meet Surface

Requirements and Member of TRB Committee AFK50 on Characteristics of Bituminous Paving Mixtures to Meet Structural Requirement, and TRB Committee AFK30 on Characteristics of Non-Asphalt Components of Asphalt Paving Mixtures. He currently serves as Flexible Pavement Section Editor of ASCE Journal of Materials in Civil Engineering and Associate Editor of the Journal of Engineering Research and International Journal of Pavement Research and Technology. He has been recognized with the 2013 Best Paper Award of the 8th International Conference on Road and Airfield Pavement Technology, 2010 Distinguished Research Paper of the Journal of Engineering Research, the 2009, 2012, and 2015 Asphalt Rubber Ambassador Award, and the 2002 Association of Asphalt Paving Technologists Board of Directors Award of Recognition.

Ragaa Abd El-Hakim, PhD Assistant Professor, Tanta University, Public Works Engineering Department, Faculty of Engineering, Tanta University, Tanta, Egypt, PH: +201008258088. E-mail: ragaa.abdelhakim@f-eng.tanta.edu.eg

Ragaa Abd El-Hakim is Assistant Professor at Tanta University, Egypt. She received a BSc degree with honor in civil engineering (structural engineering) from Tanta University, Egypt. She pursued her MSc and PhD degrees at Tanta University in civil engineering (public works engineering). She worked as Assistant Professor in Civil Engineering Department, Faculty of Engineering, Delta University for Science and Technology, from June 2010 until January 2014. She started working in Tanta University since 2014, and she was Fulbright Visiting Scholar at Texas A&M University in the academic year 2017/2018. She has taught many undergraduate and postgraduate courses. Her research interests focus on pavement material characterization and modeling, mechanistic-empirical pavement design methods, and traffic characteristics.

In-Situ Monitoring of Ground Subsidence at the Intersection of Expwy 78 and High Speed Rail of Taiwan During 2003–2011

Muhsiung Chang[1(✉)], Ren-Chung Huang[2], and Po-Kai Wu[1]

[1] Department of Civil and Construction Engineering, National Yunlin University of Science and Technology (YunTech), Yunlin, Taiwan
{changmh,wupokai}@yuntech.edu.tw
[2] Graduate School of Science and Technology, National Yunlin University of Science and Technology (YunTech), Yunlin, Taiwan
huangrope@gmail.com

Abstract. This paper discusses results of a long-term onsite monitoring on ground subsidence and soil compressibility at the intersection of Expressway (Expwy) 78 and the High-Speed Rail of Taiwan (THSR). The intersection area is located on the Chuoswei River Alluvial Fan-Delta (CRAFD), the largest and thickest alluvial deposit to the mid-west of the island. The CRAFD has been subjected to serious subsiding problems for decades because of excess extractions of groundwater for agricultural and industrial usages. The constructions of Expwy 78 and THSR in the late 1990s imposed additional loadings on the soft ground and accelerated the subsidence problem, which was becoming a threat to the safety of THSR. An 8-year onsite monitoring program at the intersection was conducted between 10/2003 and 12/2011. The subsidence and compression of soils were measured through multi-leveled magnetic rings installed in the ground along boreholes of 300-m deep, as well as a GPS station and several level-surveying benchmarks. Results indicate the ground subsidence in the intersection area was 55.7 cm for the entire deposit in the 8-year monitoring period without the loadings of Expwy 78 and THSR. The loadings of Expwy 78 embankment and THSR piers/viaducts would contribute additional subsidence of 9.4 cm and 5.5 cm, respectively, to the ground in the same period. The total subsidence in the 8-year period was 70.6 cm, with an average rate of 8.6 cm/yr. Further analysis of the compression in soils with depth <300 m indicated that the shallower deposit (depth <70 m; Aquifer F1 and Aquitard T1) was least compressible, with a strain rate of <0.01%/yr; while the deeper deposit (depth 220–300 m; Aquifers F3, F4 and Aquitards T3, T4) was most compressible, with a strain rate of 0.03–0.05%/yr. Higher compressive rates in deeper soils suggest the extraction of groundwaters has gone deeper in recent decades.

1 Introduction

Ground subsidence has been a serious issue in Taipei basin and along coastal plains of western Taiwan (Chien 1987; Wu 1987; Liao et al. 1991; Chen et al. 2007). The subsidence was due to over extraction of groundwaters for municipal usages or fishery farming. In recent decades, Chuoswei River Alluvial Fan-Delta (CRAFD), deposited by Chuoswei River to the midwest coastal plain of the island, has become a single

© Springer Nature Switzerland AG 2020
L. Mohammad and R. Abd El-Hakim (Eds.): GeoMEast 2019, SUCI, pp. 1–15, 2020.
https://doi.org/10.1007/978-3-030-34187-9_1

largest subsiding area (Hung et al. 2010). The river divides the fan-delta into Changhua County to the north and Yunlin County to the south. Previously, the area along the coastline of CRAFD suffered most serious subsidence due to exploitations of groundwater for fishery farming. The subsiding area is now moving inlands as the need of water resources for economic growth in the middle of CRAFD and the extraction of groundwaters from deeper depths.

Figure 1 indicates the current subsiding zone in Yunlin County, the southern portion of CRAFD, based on level survey data between 2003 and 2011 (WRA 2011). The accumulated subsidence has reached about 60 cm in eight years, or an average subsiding rate of 7.5 cm/yr. The subsidence has caused a serious concern on the safety of transportation structures of the area. As shown in the figure, Expressway (Expwy) 78 and Taiwan High Speed Rail (THSR) pass through the subsiding zone. Vertical alignments of these transportation arteries were distorted and threatened the safety of the structures.

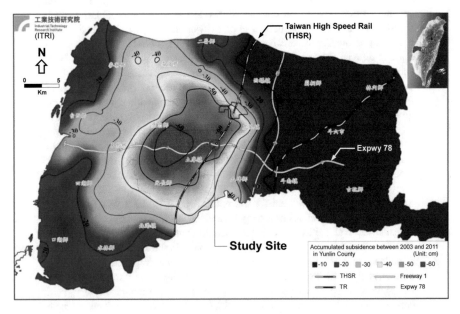

Fig. 1. Ground subsidence in Yunlin County between 2003 and 2011 (WRA 2011) and location of the study site

The aims of this study are to provide results of a long-term onsite monitoring carried between 2003 and 2011 at the intersection of Expwy 78 and THSR, as shown in Fig. 1, and to clarify the compressions of soil layers within the deposit as well as the contributions of various factors on the subsidence of the study area.

2 Background Information

The CRAFD is formed by the alluvial deposition of Chuoswei River, the longest river in Taiwan, and by the marine deposition of seawater where the sea level has been raised and lowered ±100 m several times in the recent 200 thousand years (Chappell and Shackleton 1986). In accordance, the fan-delta contains alternating layers of alluvial and marine deposits, and is considerably thick with an estimated depth of more than 350 m (Lin et al. 1992; Hung et al. 2010).

Figure 2 is a hydrogeologic model of the CRAFD. The alluvial deposits mainly consist of gravels, coarse and medium sands, and become aquifers (F-series); while the marine deposits compose of fine sand, mud and clay, and form aquitards (T-series). Based on Central Geological Survey of Taiwan (CGS 1999), four sets of aquifer/aquitard have been identified in the upper 300-m of the deposit, in which Aquifer F2 is the thickest and has been the major groundwater resources of the area.

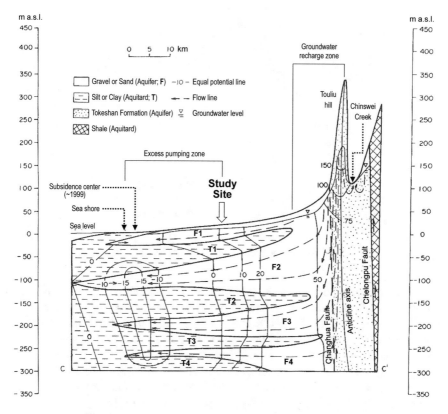

Fig. 2. Hydrogeologic model of the CRAFD (CGS, 1999)

Figure 3 indicates the layout plan of the study area. The onsite monitoring was carried out along the alignment of THSR and to the southern quadrant of the intersection of the two transportation routes. Figure 4 shows the material layer stratification of the study area based on 300-m borehole loggings at STA-1 and STA-9. The material profile indicates the deposit of the site contains interbedded layers of sandy and clayey soils, which can be further divided into four sets of aquifer and aquitard per the definitions by Central Geology Survey of Taiwan (CGS 1999).

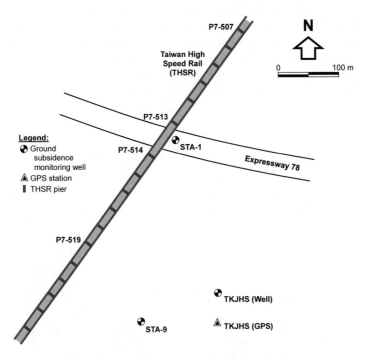

Fig. 3. Layout plan of onsite monitoring locations

Figure 5 shows the Expwy 78 and THSR of the site. Expwy 78 was built as an embankment of approximately 5.5 m high and 56 m wide, which runs in east-west direction across the CRAFD. THSR consists of a series of viaducts and piers laid in SW-NE direction and overpassed the Expwy 78 embankment in the study area. The pier foundations of THSR were formed by group piles of 2×2, 2×3 or 3×4 in arrangement, with each pile a diameter of 2 m and a length of 50–65 m.

The Expwy 78 embankment and THSR piers/viaducts of the site were constructed in different periods. As illustrated in Fig. 6, the construction of Expwy 78 embankment was carried out in three separated stages and at a start time much earlier than the 8-year monitoring period. The THSR piers/viaducts were fabricated in between Stages I and II of Expwy 78 construction, and were also completed 1.5 years prior to the onsite monitoring program discussed herein. The surcharging of Expwy 78 embankment and THSR pier foundations apparently contributed to the subsidence of the study area.

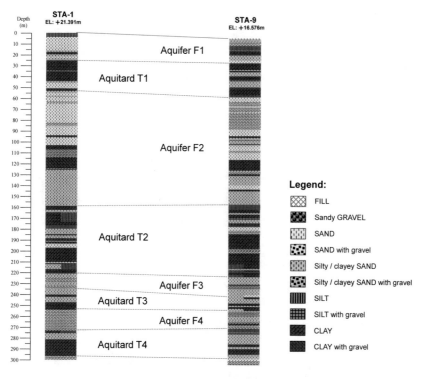

Fig. 4. Material layer stratification of the site

Fig. 5. Expwy 78 embankment and THSR piers and viaducts of the site (facing north)

Due to different construction histories, however, the contributions by Expwy 78 and THSR would not be the same in the 8-year monitoring period. To be noted, an appreciation on the complete influences of Expwy 78 and THSR on the ground subsidence could not be made based on the 8-year monitoring since both structures were constructed some times ahead of the onsite monitoring period.

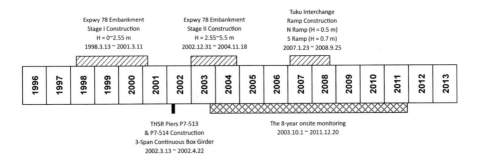

Fig. 6. Construction histories of Expwy 78 and THSR at the site

3   Ground Subsidence During 8-Year Monitoring Period

Ground subsidence behavior of the study area was observed based on various kinds of monitoring methods, including: 300-m boreholes installed with multi-leveled magnetic rings that attached to the ground (i.e., subsidence monitoring wells), a GPS station, and level survey benchmarks at the THSR piers. The monitoring of subsidence wells was conducted periodically by inserting a magnetic sensor into the well where the depths of the magnetic rings were measured and the subsidence and compressions of the ground could thus be computed. The following subsections discuss the observed subsidence as well as the contributions of various influencing factors during the 8-year monitoring period.

3.1   Summary of Different Types of Monitoring Data

Figure 7 indicates results of the subsidence monitoring well, TKJHS (well), during the 8-year monitoring period. As seen in Fig. 3, TKJHS (well) is located approximately 200 m away from Expwy 78 or THSR, as well as their intersection. The influence of the loadings from these structures on the ground subsidence at TKJHS (well) would be minimal and can therefore be neglected.

As depicted in Fig. 7, the soils within the 300-m deposit compressed progressively with time. The final subsidence of the ground at TKJHS (well) was about 40 cm in the 8-year period, or at a subsiding rate of 4.9 cm/yr, for the 300-m deposit without the influence of Expwy 78 and THSR loadings.

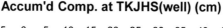

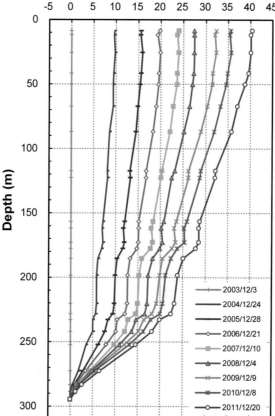

Fig. 7. Results of 8-year monitoring at TKJHS subsidence well

From the compression profile, we notice that the deeper soils with depths 220–300 m tend to compress more than does the shallower soils with depths <70 m, as depicted by the slopes of the curves. Details of the compression behaviors of the soil deposit will be discussed in the following section.

Figure 8 indicates the results of various types of monitoring at different locations. The characteristics of the monitoring data are described below:

- TKJHS (well) data – provides compressions of soils through the measurements of 300-m deep subsidence monitoring well. The monitoring location is located away from Expwy 78, THSR and their intersection (Fig. 3). The monitoring data is primarily influenced by the factors other than the loadings of Expwy 78 and THSR, i.e., nonstructural-related factors, and reflects the compressions of soils within 300 m deep.

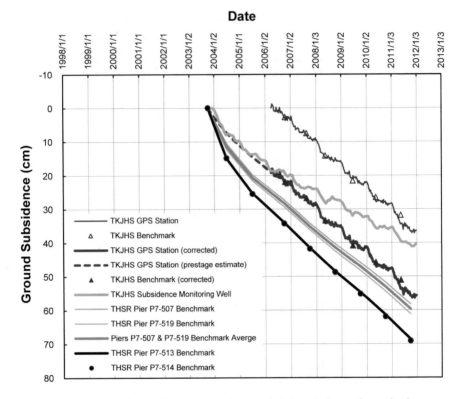

Fig. 8. Summary of different types of monitoring data during at the study site

- TKJHS (GPS) data – provides subsidence of the ground. The monitoring location is close to that of TKJHS (well) (Fig. 3). The monitoring data is also influenced by the nonstructural-related factors. However, the data reflects the compression of entire deposit, including the soils deeper than 300 m.
- TKJHS survey benchmark data – provides subsidence of the ground. The monitoring location is adjacent to TKJHS (GPS) (Fig. 3). The characteristics of the monitoring data is resembling to those of TKJHS (GPS), and reflects the compression of entire deposit, including the soils deeper than 300 m.
- THSR Piers P7-507 and P7-519 survey benchmark data – provides subsidence of the ground. The monitoring locations are along the alignment of THSR but away from Expwy 78 embankment (Fig. 3). The monitoring data is therefore influenced by THSR loading as well as nonstructural-related factors, and reflects the compression of entire deposit, including the soils deeper than 300 m.
- THSR Piers P7-513 and P7-514 survey benchmark data – provides subsidence of the ground. The monitoring locations are sitting at the intersection of Expwy 78 and THSR (Fig. 3). The monitoring data is hence influenced by the loadings of Expwy 78 and THSR, and nonstructural-related factors as well, and reflects the compression of entire deposit, including the soils deeper than 300 m.

As shown in Fig. 8, the GPS data (thin brown line) and survey benchmark data (open blue triangles) at TKJHS station agree well indicating the consistency in the measurements where the subsidence is primarily influenced by factors other than the loadings of Expwy 78 and THSR. To compare the results of subsidence monitoring well at TKJHS (well) started in late 2003 (thick green line), the TKJHS (GPS) and survey benchmark data are extrapolated based on level survey measurements of the region, as shown in Fig. 1, in approximately the same period as for the TKJHS (Well). The TKJHS (GPS) and survey benchmark data are shifted such that the ground subsidence in mid-2011 approximately equals 54 cm, as shown at the study site of the contour plot in Fig. 1.

The shifted and extrapolated TKJHS (GPS) and survey benchmark data (thick brown line and closed blue triangles) in Fig. 8 illustrate the subsidence behavior of entire deposit at the study site during the 8-year monitoring period. Compared with the data of subsidence monitoring well at TKJHS (well) (thick green line), which reflects the compression of soils within 300 m deep, we notice a substantial compression occurs in soils with depth greater than 300 m. Details of soil compressions will be discussed in the following section.

Figure 8 also shows the monitoring data at THSR pier survey benchmarks. Piers P7-507 and P7-519 are located about 200 m away from the intersection (Fig. 3), and the subsidence measured at these locations (average values in thick orange line) would b y TKJHS (GPS) (thick brown line), due to additional influence by the THSR loading. Piers P7-513 and P7-514 are situated at the intersection (Fig. 3), the subsidence measured at these locations (thick black line and closed blue dots) would also be greater than those at Piers P7-507 and P7-519 (thick orange line), due to further effect by the loading of Expwy 78.

3.2   Contributions by Various Influencing Factors

To differentiate contributions to ground subsidence by various factors, the following assumptions are made: (1) soil layers are horizontally extended; (2) groundwater level fluctuations are the same across the site; and (3) superposition principle is applicable for subsidence calculations. In accordance, the ground subsidence due to factors other than the loadings of Expwy 78 and THSR, i.e., nonstructural-related factors, can be represented by the TKJHS (GPS) data. The ground subsidence due to the loading of THSR can be estimated by subtracting the TKJHS (GPS) data from the level survey benchmark data at THSR Piers P7-507 and P7-519. Similarly, the ground subsidence due to the loading of Expwy 78 can be assessed by subtracting the level survey data at THSR Piers P7-507 and P7-519 from the level survey data at THSR Piers P7-513 and P7-514.

Figure 9 presents results of the above assessment on the contributions of various factors to the ground subsidence of the site. Table 1 indicates the contributions to the ground subsidence during the 8-year monitoring period are 13% and 8%, respectively, by the loadings of Expwy 78 and THSR. Although Expwy 78 embankment appears to be more influential than THSR piers/viaducts, their contributions to ground subsidence are significantly less than those by nonstructural-related factors.

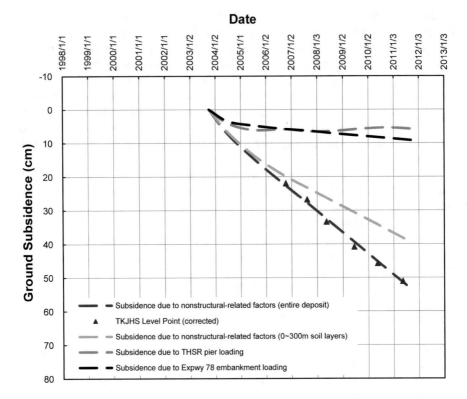

Fig. 9. Influences of various factors on the subsidence behavior of the site

Table 1. Contributions of various factors on the subsidence of the site during the 8-year monitoring period (2003.10.1– 2011.12.20)

Influence factor	Subsidence contribution		
	(cm)	(%)	(cm/yr)
Nonstructural-related factors: groundwater fluctuation, overpumping, soil creeping, etc.	55.7	79	6.78
Expwy 78 embankment loading	9.4	13	1.14
THSR pier loading	5.5	8	0.67
Total	70.6	100	8.59

The nonstructural-related factors in a broader sense would include the short-term fluctuations and long-term declines of groundwater levels due to infiltration or evaporation of rainfalls or surface waters, or due to human extraction of groundwaters, as well as the creeping of soil deposit due to its great amount of thickness. As indicated previously, the subsiding area has been moving inlands of the CRAFD in recent decades due to economic growth of the areas and extractions of groundwater from deeper depths. Apparently, the drops in groundwater levels by overpumping would have a significant contribution on the subsidence of the site. Besides, the CRAFD consists of soil deposits of more than 350 m in thickness (Lin et al. 1992; Hung et al. 2010). In spite the soils might have completed their normal consolidation process, insignificant creeping in soils can still accumulate deformations with time to an amount that cannot be neglected.

It should be noted, however, the contributions of various factors on the subsidence of the site cannot be fully addressed based solely on the results of 8-year monitoring. As pointed in Fig. 6, the constructions of Expwy 78 embankment and THSR piers and viaducts consisted of separated stages and started prior to the 8-year monitoring. The influences of Expwy 78 and THSR loadings would be decreasing with time. In the 8-year monitoring period, the contributions of Expwy 78 and THSR loadings to the subsidence were apparent in the first two years, and then became stabilized as shown in Fig. 9.

4 Soil Compression During 8-Year Monitoring Period

This section discusses compressions and compressibility of soil layers based on the monitoring data of magnetic rings installed in the subsidence well TKJHS (well).

4.1 Compressions of Soils at Different Depth Intervals

As shown previously in Fig. 7, the compression profile at TKJHS (well) can be divided into three depth intervals, of which the calculated compressions and compressive strains are indicated in Table 2. In the shallower depth of less than 70 m, the compression would be the least, accounted for 6% of the total subsidence in the 8 years, and the compressibility would be the smallest, with a strain rate of 0.047 cm/m. For the deeper depth range of 220–300 m, however, the compression would be the greatest, accounted for 42% of the total subsidence in the 8 years, and the compressibility would be the largest, with a strain rate of 0.290 cm/m.

Table 2. Compressions of soil at different depth intervals of the deposit during the 8-year monitoring period (2003.10.1–2011.12.20)

Depth interval	Compression & compressive strain		
	(cm)	(%)	(cm/m)
0–70 m	3.3	6	0.047
70–220 m	13.9	25	0.093
220–300 m	23.2	42	0.290
>300 m	15.3	27	–
Total	55.7	100	–

The depth range of 220–300 m comprises aquifers F3 & F4 and aquitards T3 & T4. Since Aquifer F2 (depth range 55–160 m, approx.) used to be the major groundwater resource of the area, the observed greater compressions and compressibility of deeper layers suggest the extraction of groundwaters should have gone deeper in recent decades.

Table 2 also indicates the compression of soils with depth deeper than 300 m, based on the difference of monitoring data between TKJHS (GPS) and TKJHS (well), as depicted in Fig. 8. Results indicate the compression of soils deeper than 300 m (i.e., the monitoring depth of subsidence well TKJHS (well)) would be substantial, accounted for 27% of the total subsidence in the 8 years.

In view of the installation depth of THSR group piles of 50–65 m, the compression of shallower soils (i.e., 3.3 cm in 8 years; for depth <70 m) appears small and would not cause adverse effects (i.e., negative skin frictions) on the piles. The compression of deeper soils (i.e., 52.4 cm in 8 years; for depth >70 m), however, is substantial and would be detrimental to the vertical alignment of the THSR structure.

4.2 Compressions of Aquifer and Aquitard Layers

Compressions of soils are further analyzed in terms of aquifer and aquitard layers, and results shown in Fig. 10 and Table 3. As illustrated in Fig. 10(a), Aquifer F1 and Aquitard T1 presented least compressions, while Aquifers F2 & F4 experienced most compressions, over the 8-year monitoring period.

However, the amount of compression would be affected by the layer thickness. The compressive strain or compressive strain rate is adopted instead. As shown in Fig. 10(b) and Table 3, Aquifer F1 and Aquitard T1 have least compressibility, or a compressive strain rate of <0.01%/yr; while Aquifers F3 & F4 and Aquitards T3 & T4 exhibit greatest compressibility, or a compressive strain rate of 0.03–0.05%/yr.

Aquifer F1 and Aquitard T1 of the site are situated at a depth of <60 m. The minimal compressibility of the layers might suggest the associated soils are in a slightly overconsolidated (OC) condition. Aquifers F3 & F4 and Aquitards T3 & T4 are located at a depth range of 220–300 m. The greater compressibility of these layers, however, indicates the associated soils are approximately in a normally consolidated (NC) state. Greater compressibility of deeper soils might also suggest the extraction of groundwaters of the study area had gone deeper in the recent decades.

Date

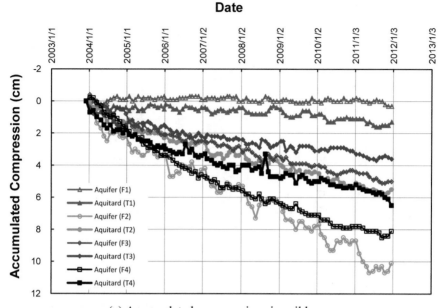

(a) Accumulated compressions in soil layers

Date

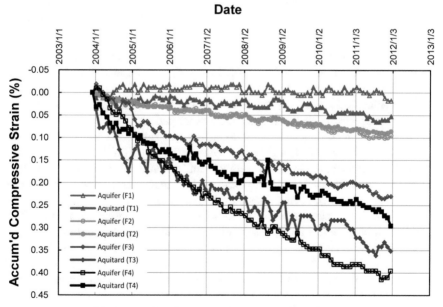

(b) Accumulated compressive strains in soil layers

Fig. 10. Compressions and compressive strains of soil layers at the site

Table 3. Compressions of aquifers and aquitards measured at TKJHS Station during the 8-year monitoring period (2003.10.1–2011.12.20)

Soil layer	Thickness	Compression	Compression rate	Compressive strain rate
	(m)	(cm)	(cm/yr)	(%/yr)
Aquifer F1	16.83	0.3	0.04	0.002
Aquitard T1	24.70	1.3	0.16	0.007
Aquifer F2	106.28	10.1	1.25	0.012
Aquitard T2	63.75	5.5	0.68	0.011
Aquifer F3	21.70	5.0	0.62	0.029
Aquitard T3	10.23	3.6	0.44	0.044
Aquifer F4	20.46	8.1	1.00	0.049
Aquitard T4	21.97	6.5	0.80	0.037
Total	285.92	40.4	4.98	0.018

5   Conclusions

This paper discusses results of a long-term monitoring carried between 2003 and 2011, an 8-year onsite monitoring at the intersection of Expwy 78 and THSR, with the aims to clarify the compression behavior of soils in the CRAFD and the contributions of various factors on the subsidence of the study area. Major findings of this study are listed as follows:

- The ground subsidence in the intersection area is 55.7 cm for the entire deposit in the 8-year monitoring period without the loadings of Expwy 78 and THSR, i.e., structural-related factors.
- The loadings of Expwy 78 and THSR would contribute additional 9.4 cm and 5.5 cm, respectively, to the ground in the same period.
- Nonstructural-related factors appear dominating the subsidence behavior of the site. Nonstructural-related factors generally include the short-term fluctuations and long-term declines in groundwater levels due to infiltration or evaporation of rainfalls or surface waters, or due to human extraction of groundwaters, as well as the creeping of soil deposit which is enormously thick (>350 m).
- To be noted, the contributions of various factors on the subsidence cannot be fully addressed based solely on the results of 8-year monitoring. In viewing that Expwy 78 embankment and THSR piers/viaducts were constructed in separated stages and started prior to the 8-year monitoring, the influences by Expwy 78 and THSR loadings would only be partially reflected in the period considered in this study.
- The monitored compressions in soils reveal the shallower deposit (depth <70 m; Aquifer F1 and Aquitard T1) is least compressible, with a strain rate of <0.01%/yr; while the deeper deposit (depth ranged 220–300 m; Aquifers F3 & F4 and Aquitards T3 & T4) is most compressible, with a strain rate of 0.03–0.05%/yr.
- The compression of shallower depths (<70 m; 3.3 cm in 8 years) appears small and would not cause adverse effects (i.e., negative skin frictions) on THSR piles.

However, the compression of deeper depths (>70 m; 52.4 cm in 8 years) is substantial and would be detrimental to the vertical alignment of THSR structures.

• Higher compressibility of deeper soils suggests the extraction of groundwaters in the study area has gone deeper in the recent decades.

Acknowledgments. The authors would like to thank the financial supports by Resources Engineering, Inc. Taiwan (NYST 102-272) and Ministry of Science and Technology (previously, National Science Council), Taiwan (NSC102-2815-C-224-020-E, MOST105-2815-C-224-003-E). Some background information and monitoring data provided by Central Geological Survey, Water Resources Agency, and Directorial General of Highways, Taiwan, are gratefully appreciated. The authors would like to thank the following personnel, J. H. Jhang, Y. C. Chang, C. F. Chuang and other members of Geotechnical Engineering Team of YunTech for carrying out the onsite monitoring work.

References

Central Geological Survey (CGS): Summary report of groundwater monitoring network plan in Taiwan - Phase I (1992–1998), Ministry of Economic Affairs, Taiwan (1999). (in Chinese)

Chappell, J., Shackleton, N.J.: Oxygen isotopes and sea level. Nature **324**, 137–140 (1986). https://doi.org/10.1038/324137a0

Chen, C.T., Hu, J.C., Lu, C.Y., Lee, J.C., Chan, Y.C.: Thirty-year land elevation change from subsidence to uplift following the termination of groundwater pumping and its geological implication in the Metropolitan Taipei Basin, Northern Taiwan. Eng. Geol. **95**, 30–47 (2007). https://doi.org/10.1016/j.enggeo.2007.09.001

Chien, J.Y.: Ground subsidence problem along coastal areas of Taiwan. Sino-Geotech. **20**, 50–56 (1987). (in Chinese)

Hung, W.C., Hwang, C., Chang, C.P., Yen, J.Y., Liu, C.H., Yang, W.H.: Monitoring severe aquifer-system compaction and land subsidence in Taiwan using multiple sensors: Yunlin, the southern Choushui River Alluvial Fan. Environ. Earth Sci. **59**, 1535–1548 (2010). https://doi.org/10.1007/s12665-009-0139-9

Liao, J.S., Pan, K.L., Haimson, B.C.: The monitoring and investigation of ground subsidence in southwest Taiwan. In: Proceedings of 4th International Symposium on Land Subsidence, pp. 81–96. IAHS Publication 200, May 1991

Lin, L.H., Lin, H.R., Ke, A.H.W., Chou, T.H.: Petroleum potential of the pre-miocene formations in the Chianan Plain, Taiwan. Petrol. Geol. Taiwan **27**, 177–197 (1992)

Water Resources Agency (WRA): Multi-sensors Applied to Monitor Subsidence and Investigate Mechanism in Taipei, Changhua and Yunlin Area in 2011, Ministry of Economic Affairs, Taiwan (2011). ISBN 978-986-03-0205-9. (in Chinese with English abstract)

Wu, J.M.: A retrospect on ground subsidence of Taipei Basin. Sino-Geotech. (20), 34–49 (1987). (in Chinese)

Improving Performance of Testing Laboratories – A Statistical Review and Evaluation

G. I. Anastasopoulos[1]([✉]), P. S. Ramakrishnan[1], and I. G. Anastasopoulos[2]

[1] International Accreditation Service, Brea, USA
ganas@checkp.info
[2] School of Letters and Sciences, University of California at Berkeley, Berkeley, USA

Abstract. This paper outlines the most common quality challenges testing laboratories are facing during their accreditation process. Accreditation is the independent evaluation of conformity assessment bodies (i.e. Testing laboratories) against recognized standards to carry out specific activities to ensure their impartiality and competence. Through the application of national and international standards, government, procurers and consumers can have confidence in the quality of test results, inspection reports and certifications provided.

This study has been performed based on data collected by more than 300 testing laboratories, from 41 countries worldwide, accredited against the requirements of the international standard ISO/IEC 17025-2005 *"General requirements for the competence of testing and calibration laboratories."*

The Non-Conformities were issued during the accreditation process of various testing laboratories specializing in different testing categories (Civil, Geotechnical, Mechanical, Electrical, Chemical, Microbiological, etc.). Findings vary from commonly reported quality management system issues to the most demanding technical challenges faced by testing laboratories.

The identified Non-Conformities were categorized and statistically processed. The trends are identified and analyzed per quality management system or technical category. Under the accreditation process, laboratories are required to respond to any significant findings with a submittal of a corrective action plan containing an analysis of the root cause, details of actions taken to resolve the issue and strategies to prevent reoccurrence. Various responses were analyzed and some suggestions and good practices were gleaned from these submittals. Opportunities for improvement are presented for each corresponding category of findings.

ISO/IEC 17025:2005 specifies the general requirements for the competence to carry out tests and/or calibrations, including sampling. It covers testing and calibration performed using standard methods, non-standard methods, and laboratory-developed methods. It is applicable to all organizations performing tests and/or calibrations. These include, for example, first-, second- and third-party laboratories, and laboratories where testing and/or calibration forms part of inspection and product certification. The standard is applicable to all laboratories regardless of the number of personnel or the extent of the scope of testing and/or calibration activities.

© Springer Nature Switzerland AG 2020
L. Mohammad and R. Abd El-Hakim (Eds.): GeoMEast 2019, SUCI, pp. 16–34, 2020.
https://doi.org/10.1007/978-3-030-34187-9_2

1   General

ISO/IEC 17025 was first issued in 1999 by the International Organization for Standardization (ISO) and the International Electro-technical Commission (IEC). It is the single most important standard for calibration and testing laboratories around the world. CASCO is the ISO committee that works on issues relating to conformity assessment. CASCO develops policy and publishes standards related to conformity assessment. CASCO's standards development activities are carried out by working groups made up of experts put forward by the ISO member bodies. The experts are individuals who possess specific knowledge relating to the activities to be undertaken by the working group [1].

The data provided in this paper refers to the ISO/IEC 17025 version 2005 [2]. Since December 2017 the ISO/IEC 17025 version 2017 [3] of the standard is available, with a transition period until December 2020. After that date the 2005 version will not be used anymore and it will be replaced by the 2017 version of the standard. In order to facilitate the reader, at the end of this paper there is a table (see Annex) with corresponding clauses of 2005 and 2017 versions of the standard.

At the International Laboratory Accreditation Cooperation (ILAC) General Assembly in October 2013 the Laboratory Committee (which is composed of stakeholder representatives of accredited testing and calibration) recommended that ILAC request that ISO/CASCO establish a new work item to comprehensively revise ISO/IEC 17025:2005. The 6th ISO/CASCO WG 44 meeting was held on July 10–12, 2017 in ISO Central Secretariat, Geneva. The deliverable of this meeting was the FDIS version of the new ISO/IEC 17025 version. The document was published at November 2017.

Please note that throughout this article the term "the standard" refers to the new ISO/IEC 17025:2005.

2   Scope

According to ISO [4] and ILAC [5] more than 68.000 calibration as well as testing laboratories, worldwide, are accredited to ISO/IEC 17025 standard, from more than 120 Accreditation Bodies, out of which 98 are ILAC MRA signatories. The ILAC Mutual Recognition Arrangement (ILAC MRA) provides significant technical underpinning to the calibration, testing, medical testing and inspection results of the accredited conformity assessment bodies and in turn delivers confidence in the acceptance of results. The ILAC MRA enhances the acceptance of products across national borders. By removing the need for additional calibration, testing, medical testing and/or inspection of imports and exports, technical barriers to trade are reduced. In this way the ILAC MRA promotes international trade and the free-trade goal of "accredited once, accepted everywhere" can be realized [6].

Testing Laboratories are using ISO/IEC 17025 standard to implement a quality system aimed at improving their ability to consistently produce valid results. Since the standard is about competence, accreditation is simply a formal recognition of a demonstration of that competence. A prerequisite for a laboratory to become accredited is to have a documented quality management system. Regular internal audits are expected to indicate opportunities to make the test or calibration better than it was. The laboratory is also expected to keep abreast of scientific and technological advances in relevant areas.

This study has been performed based on data collected by more than 300 testing laboratories, from 41 countries worldwide, accredited against the requirements of the international standard ISO/IEC 17025:2005 *"General requirements for the competence of testing and calibration laboratories."*

The Non-Conformities were issued during the accreditation process of various testing laboratories specializing in different testing categories (Mechanical, Electrical, Geotechnical, Chemical, Microbiological, etc.). Findings vary from commonly reported management system issues to the most demanding technical challenges faced by testing laboratories.

The identified Non-Conformities were categorized and statistically processed. The trends are identified and analyzed per management system or technical category. Under the accreditation process, testing laboratories are required to respond to any significant findings with a submittal of a corrective action plan containing an analysis of the root cause, details of actions taken to resolve the issue and strategies to prevent reoccurrence [7]. Various responses were analyzed and suggestions and good practices were gleaned from these submittals. Opportunities for improvement are presented for each corresponding category of findings.

3   Analysis – Results

The non-conformance analysis of data was performed across all countries. Here is the breakdown of number of laboratories per country that was selected for data analysis (Fig. 1).

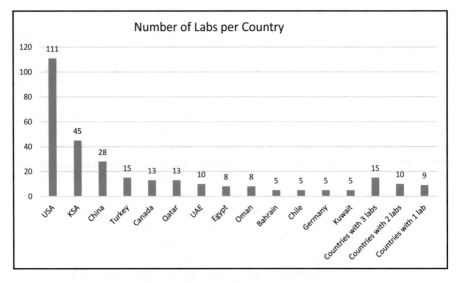

Fig. 1. Number of Labs per Country

Five major fields (scopes of accreditation) were identified and grouped for this analysis:

- Geotechnical
- Mechanical/Physical/Structural
- Electrical/Electronics
- Chemical/Microbiology/Environmental
- Medical

At Fig. 2, below, is the pictorial representation of number of laboratories per scope and number of laboratories identified for this analysis:

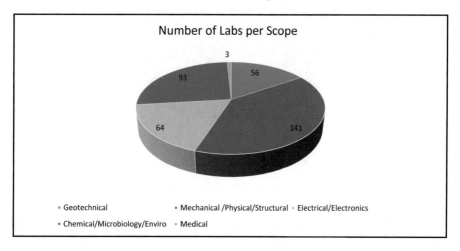

Fig. 2. Number of Labs per Scope

At Fig. 3, below, is the pictorial representation of the number of laboratories per scope and number of laboratories identified of the top five countries with the largest volume of laboratories for this analysis:

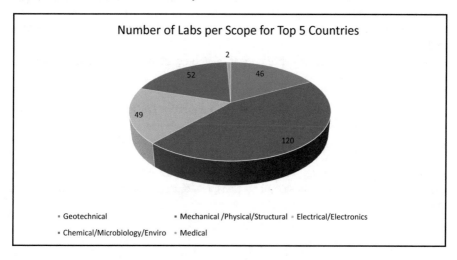

Fig. 3. Number of Labs per Scope for top 5 countries

A total of 1510 non-conformities noted during the assessments that were reviewed. Based on the analysis, the distribution of non-conformities was measured as described in the Fig. 4, below:

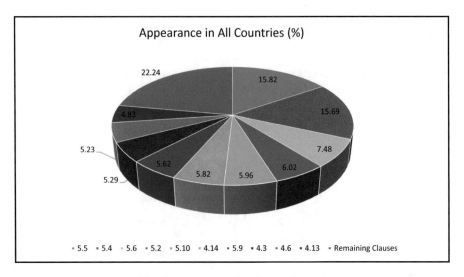

Fig. 4. Appearance in all countries (%)

Upon observation of the chart above, it is noted that the most common of the non-conformities were cited on clauses 5.5 (Equipment) and 5.4 (Test and calibration methods).

The group of clauses 5.6 (Measurement traceability), 5.2 (Personnel), 5.10 (Reporting the results), 4.14 (Internal audits), 5.9 (Assuring the quality of test and calibration results), 4.3 (Document control), 4.6 (Purchasing services and supplies) and 4.13 (Control of records) formed the next layer of non-conformities with similar percentages of findings.

Finally, a series of other clauses were minor ones constituting 22.24% of the non-conformities in the collected data.

The top ten clauses on which the most of non-conformities were cited were:

1. 5.5-Equipment.

This clause refers to the policies and procedures for ensuring equipment used for testing are available, suitable and properly maintained.

2. 5.4-Test and calibration methods.

This clause refers to the policies and procedures for choosing methods of testing and calibration (which covers sampling, transport, storage, uncertainty, control of data etc.).

3. 5.6-Measurement traceability.

This clause refers to the procedure for choosing, using, calibrating, checking and maintaining measurement standards, reference materials used as measurement standards, and equipment used for testing.

4. 5.2-Personnel.

This clause refers to the measures taken to ensure that all laboratory staff is properly skilled and qualified.

5. 5.10-Reporting the results.

This clause refers to the measures taken to ensure that results of testing are reported clearly and objectively.

6. 4.14-Internal audits.

This clause refers to the policies and procedures for conducting internal audits and implementing findings.

7. 5.9-Assuring the quality of test and calibration results.

This clause refers to the procedures for monitoring the validity of testing.

8. 4.3-Document control.

This clause refers to the procedures for:

A. Controlling all documents (internal and external) relating to the QMS – regulations, normative reference documents, drawings, specifications, instructions, manuals etc.
B. Approving and issuing documents (including maintaining a master list).
C. Changing/correcting documents.

9. 4.6-Purchasing services and supplies.

This clause refers to the policies and procedures for choosing and buying services and supplies that, when used, may affect the quality of tests.

10. 4.13-Control of records.

This clause refers to the procedures for controlling records (identification, collection, indexing, access, filling, storage, maintenance and disposal of quality and technical records).

The clauses referred above and all the data provided in this paper refers to the ISO/IEC 17025 version 2005 [2]. Since December 2017 the ISO/IEC 17025 version 2017 [3] of the standard is available, with a transition period until December 2020. After that date the 2005 version will not be used anymore and it will be replaced by the 2017 version of the standard. In order to facilitate the reader, at the end of this paper there is a table (see Annex) with corresponding clauses of 2005 and 2017 versions of the standard.

This standard was developed with the objective of promoting confidence in the operation of laboratories and contains requirements for laboratories to enable them to demonstrate that they operate in a competent and impartial way and that they are able to provide valid results.

It is important to be noted that the new update to ISO/IEC 17025:2017 is introducing greater emphasis on the responsibilities of senior management, risk analysis, impartiality and explicit requirements for continual improvement of the management system itself, and particularly, communication with the customer. Laboratories that use ISO/IEC 17025, version 2005, that have not demonstrated full compliance with new ISO/IEC 17025:2017 international standard by December 1, 2020, are subject to suspension/cancellation of their accreditation status.

Analysis of these top ten clauses across the top five countries with the highest laboratories population was performed and similar trends were observed as presented in Fig. 5, below:

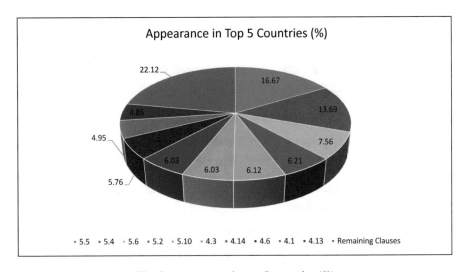

Fig. 5. Appearance in top 5 countries (%)

The top ten clauses on which most non-conformities were cited for the Top 5 Countries, with higher number of laboratories, were:

1. 5.5 (Equipment)
2. 5.4 (Test and calibration methods)
3. 5.6 (Measurement traceability)
4. 5.2 (Personnel)
5. 5.10 (Reporting the results)
6. 4.3 (Document control)
7. 4.14 (Internal audits)
8. 4.6 (Purchasing services and supplies)
9. 4.1 (Organization)
10. 4.13 (Control of records).

The trends between all countries when compared to the ones from the top 5 countries in laboratories population are very similar. It is concluded that the distribution of non-conformities is consistent among the countries with many accredited labs and the ones with less. This is a result of:

- The globalized approach on accreditation rules and guidelines as issued by ILAC and regional ILAC members.
- The harmonized approach on management system documentation in global level, followed by consultants.
- The more homogeneous training programs, facilitating a uniform approach in the design and the implementation of testing laboratory accreditation systems.

A detailed analysis of non-conformities across various scopes of accreditation was also performed and based on that analysis, the following trends were observed (Fig. 6):

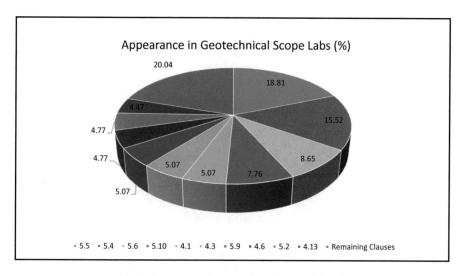

Fig. 6. Appearance in geotechnical scope labs (%)

In the geo-technical scope, most of the non-conformities were cited on the following clauses:

1. 5.5 (Equipment)
2. 5.4 (Test and calibration methods)
3. 5.6 (Measurement traceability)
4. 5.10 (Reporting the results)
5. 4.1 (Organization)
6. 4.3 (Document control)
7. 5.9 (Assuring the quality of test and calibration results)
8. 4.6 (Purchasing)
9. 5.2 (Personnel) and
10. 4.13 (Control of records)

The trends among geo-technical scope are typical to the ones observed among all laboratories. The distribution of non-conformities among labs follows the usual trends. It is interesting though to note the following:

- Technical issues (5.5-Equipment, 5.4-Test and calibration methods, 5.6-Measurement traceability and 5.10-Reporting) produce the majority of non-conformities. Those results remain on the top 4 and constitute more than half of the raised non-conformities.
- Clause 4.1 (Organization) is relatively much higher compared to other types of laboratories. This can be explained due to the nature of geo-technical laboratories, the extended external work performed and the operation of many site laboratories.

The distribution of the non-conformities for mechanical/physical/structural scope was also performed and based on that analysis, the following trends were observed (Fig. 7):

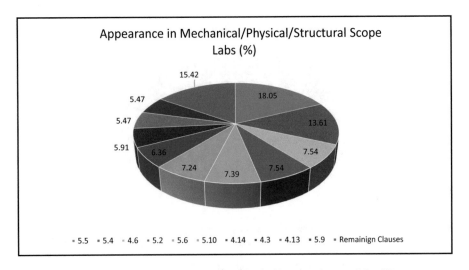

Fig. 7. Appearance in mechanical/physical/structural scope labs (%)

In the mechanical/physical/structural scope, most of the non-conformities were cited on the following clauses:

1. 5.5 (Equipment)
2. 5.4 (Test and calibration methods)
3. 4.6 (Purchasing)
4. 5.2 (Personnel)
5. 5.6 (measurement traceability)
6. 5.10 (Reporting of results)
7. 4.14 (Internal audits)
8. 4.3 (Document control)
9. 4.13 (Control of records) and
10. 5.9 (Assuring the quality of test and calibration results).

The trends among mechanical, physical and structural scope are also typical to the ones observed among all laboratories. The distribution of non-conformities among labs follows the usual trends. It is interesting though to note the following:

- Technical issues (5.5-Equipment, 5.4-Test and calibration methods, 5.6-Measurement traceability, 5.2-Personnel and 5.10-Reporting of results) produce the majority of non-conformities. Those results remain on the top 6 and constitute more than half of the raised non-conformities.
- Clause 4.6 (Purchasing) is relatively much higher (3rd) compared to other types of laboratories. This can be explained as luck of interesting in performing such activities, probably due to sufficient number of complying suppliers.

The distribution of the non-conformities for electrical/electronic was also performed and based on that analysis, the following trends were observed (Fig. 8):

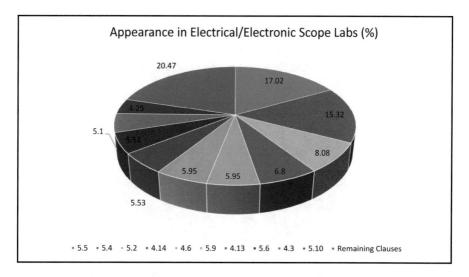

Fig. 8. Appearance in electrical/electronic scope labs (%)

In the electrical/electronic scope, most of the non-conformities were cited on the following clauses:

1. 5.5 (Equipment)
2. 5.4 (Test and calibration methods)
3. 5.2 (Personnel)
4. 4.14 (Internal audits)
5. 4.6 (Purchasing)
6. 5.9 (Assuring the quality of test and calibration results)
7. 4.13 (control of records)
8. 5.6 (Measurement traceability)
9. 4.3 (Document control) and
10. 5.10 (Reporting of results)

The trends among electrical and electronic scope are also typical to the ones observed among all laboratories. The distribution of non-conformities among labs follows similar trends. It is interesting though to note the following:

- Clause 5.10 (Reporting of results) is ranked last at the list of first 10, similarly to 5.6 (measurement traceability) which is no. 10. This indicates a relatively higher level of technical compliance than the rest of the sample.
- Clause 4.14 (Internal audits) is relatively ranked higher (no. 4) than the average of the sample. Similarly clause 4.6 (Purchasing) is relatively higher (5rd) compared to other types of laboratories. This can be explained as luck of interesting in performing such activities, probably due to sufficient number of complying suppliers.

The distribution of the non-conformities for Chemical/Microbiology/ Environmental was also performed and based on that analysis, the following trends were observed (Fig. 9):

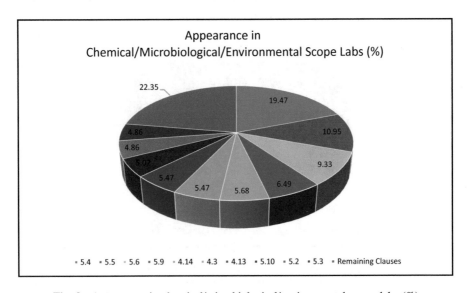

Fig. 9. Appearance in chemical/microbiological/environmental scope labs (%)

In the Chemical/Microbiology/Environmental scope, most of the non-conformities were cited on the following clauses:

1. 5.4 (Test methods)
2. 5.5 (Equipment)
3. 5.6 (Measurement traceability)
4. 5.9 (Assuring the quality of test and calibration results)
5. 4.14 (Internal audits)
6. 4.3 (Document control)

7. 4.13 (Control of records)
8. 5.10 (Test reports and calibration certificates)
9. 5.2 (Personnel) and
10. 5.3 (Accommodation and environmental conditions)

The trends among different fields/scopes, when compared to each other, are also very similar. It is concluded that the distribution of non-conformities among labs of different fields follows the same trends. It is interesting to note the following:

- Clauses 5.4 (Test methods) and clauses 5.5 (Equipment) remain the top sources of non-conformities on all types of testing laboratories.
- Clause 4.14 (Internal audits) is not the top, but it remains a repeated cause of non-conformities throughout all types of testing laboratories.
- Clause 5.3 (Accommodation and environmental conditions) is more common cause of non-conformities in the Chemical/Microbiology/Environmental than the other ones, due to the nature of the tests.

The next task included analysis of data performed in a region-wise approach. The analysis results are presented below (Fig. 10):

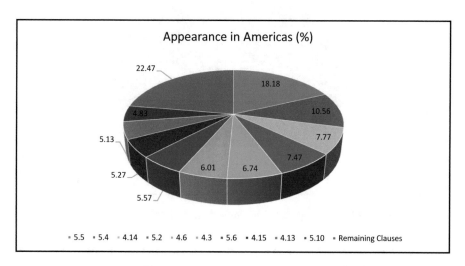

Fig. 10. Appearance in Americas (%)

In the region of North, Central and South America, most of the non-conformities were cited on the following clauses:

1. 5.5 (Equipment)
2. 5.4 (Test and calibration methods)
3. 4.14 (Internal audits)
4. 5.2 (Personnel)
5. 4.6 (Purchasing)
6. 4.3 (Document control)

7. 5.6 (Measurement traceability)
8. 4.15 (Management review)
9. 4.13 (Control of records) and
10. 5.10 (reporting the results)

The trends of North, Central and South America are indicative of the following:

- Clause 5.10 (reporting the results) is last on the list of non-conformities, which makes it the strong point for American labs. The mature market, the litigation risks and the advanced document processing systems are the main drivers for this fact.
- Clause 4.14 (Internal audits) is high for this region. With 4.15 (Management review) they are reaching close to 14% of non-conformities.

The distribution of the non-conformities for Europe, Middle East and Africa was performed and based on that analysis, the following trends were observed (Fig. 11):

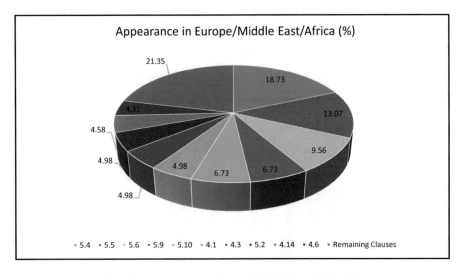

Fig. 11. Appearance in Europe/Middle East/Africa (%)

In the region of Europe, Middle East and Africa, most of the non-conformities were cited on the following clauses:

1. 5.4 (Test and calibration methods)
2. 5.5 (Equipment)
3. 5.6 (Measurement traceability)
4. 5.9 (Assuring the quality of test and calibration results)
5. 5.10 (reporting the results)
6. 4.1 (Organization)
7. 5.3 (Accommodation and environmental conditions)
8. 5.2 (Personnel)
9. 4.14 (Internal audits)
10. 4.6 (Purchasing)

The trends of Europe, Middle East and Africa are characteristic of:

- Strong weakness in evident in Technical issues in that geographical region. Since the majority of the sample comes from Middle-East and Gulf countries, it is evident that the usage of many ex-patriot technicians and the high rates of changing personnel are negatively affecting the lab operation.
- Clause 4.14 (Internal audits) is the strong point for this region. With 4.6 (Purchasing) they are reaching less than 10% of non-conformities.

The distribution of the non-conformities for Asia was also performed and based on that analysis, the following trends were observed (Fig. 12):

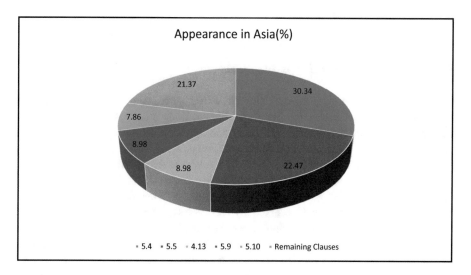

Fig. 12. Appearance in Asia (%)

In the region of Asia, most of the non-conformities were cited on the following clauses:

1. 5.4 (Test and calibration methods)
2. 5.5 (Equipment)
3. 4.13 (Control of records)
4. 5.9 (Assuring the quality of test and calibration results) and
5. 5.10 (Reporting the results)

The trends of Asia are characteristic of:

- Strong weakness in Technical issues mainly clauses 5.4 (Test and calibration methods) and 5.5 (Equipment) that constitute more than 50% of the non-conformities.
- Clause 4.13 (Control of records) is also high (3rd) a little less than 10%.

- Asian testing laboratories tend to underestimate the value of a well prepared test report and many times they don't take under consideration the test report requirements of the individual testing standards. Relative clause 5.10 (Reporting the results).

The overall trends among all different regions can be summarized as follows:

- Test and calibration methods and Equipment are remaining the main sources of findings.
- Internal Auditing can be considered a significant problem in the American region.
- Traceability remains a significant problem in the Europe, Middle East and Africa regions.
- Asian region is weak in Control of Records and Testing Reports.

4   Opportunities for Improvement

Usually, at the end of each accreditation (or surveillance, or re-accreditation) assessment, during the closing meeting, the Accreditation Body's Lead Assessor will present the findings and summarize whether the Laboratory's operation is in conformance or not to Standard and Management System requirements. If not fully in conformance, the Assessor will work with the management of the laboratory in order to develop a time line of corrective actions. Satisfactory proof of acceptable corrective actions should be then submitted by the laboratory to Accreditation Body. When the submitted corrective actions are implemented and accepted by the Assessor, then the laboratory can be accredited, or retain its accreditation in the case that it has been already accredited.

In addition to accreditation assessment, most accreditation standards/criteria, including ISO/IEC 17025 require from the laboratory to perform internal audits on a regular basis. Continual improvement is basic element of most management system standards so the continuing effectiveness of the laboratory's management system is a key issue.

Based on the corrective actions submitted by the assessed laboratories, in response to identified non-compliances, described in details in our present analysis, a series of opportunities for improvement have been identified and implemented. We are summarizing the main opportunities for improvement below:

1. Laboratories are advised to carefully implement the nationally/internationally recognized test methods.
2. In case, any test method is developed by the laboratory, it needs to be validated. Laboratories should modify standard methods or develop their own method, only if it is a requirement. In that case appropriate validation records must be prepared and provided
3. Laboratories need to ensure that they employ competent personnel, capable to perform measurement uncertainty understanding and being in position to explain the theory and mechanisms behind it. It is advised to train those personnel in specialized courses explaining the details of the uncertainty of measurement.

4. Testing and other related data needs to be checked and transferred carefully. The laboratory should be careful with securing control of data.
5. With regard to equipment, it is suggested that laboratory keeps the equipment inventory with all the relevant information such as calibration certificates, maintenance records and manuals in place where it can be easily accessible.
6. Maintenance should be performed in time frames described in a preventive maintenance schedule, and appropriate records should be available to trace any past problems and actions.
7. The laboratories can keep an automated system in their Laboratory Information Management Software (LIMS) reminding them on the upcoming calibration schedules well in advance to avoid the unnecessary delays.
8. The laboratories need to keep the unbroken chain of tracing the measurements to relevant primary standards of measurement standards. This is usually possible by calibrating their measuring devices at ISO/IEC 17025 accredited calibration laboratories. Attention should be paid that the calibrated instruments are covered by the scope of accreditation of the ISO/IEC 17025 accredited calibration laboratories.
9. When using reference standards, the laboratory need to keep a schedule for the calibration of reference standards and a maintenance plan for the same.
10. When using certified reference materials, laboratory staff needs to know the key parameters and the effective use of the reference materials.
11. Laboratories need to lay out a plan for internal audits and management reviews to do it periodically (recommended annually) and maintain effective documentation.
12. Please note that, while performing fixes to the findings noted during the internal/external audits, the laboratory needs to perform root cause analysis and come up with an effective action to fix it from recurring. There is distinction between correction and corrective action. The Standard is asking for corrective actions and not corrections.
13. With regard to purchasing, it is suggested to perform periodic review of the approved vendors/suppliers of critical consumables on a routine basis. Records of this evaluation/review are expected to be available.
14. With regard to assuring quality of test results, laboratories need to identify possible participation in proficiency testing, inter-lab comparison besides performing replication of tests. Other assuring quality of test results can be acceptable such as replication tests, repeatability tests, comparison of test results performed by different technicians etc.
15. A plan/schedule with regards to assuring quality of test results, as described in the point above, is expected. The plan can be from one to four years (a full accreditation cycle). It is expected that all tests under the scope of accreditation ac covered under this plan.
16. Laboratories need to pay attention to the reporting requirements of the ISO/IEC 17025 standard and appropriate technical standards using which the tests are performed. It is important to note that in addition to ISO/IEC 17025 many testing standards are also requiring specific information to be included in the test report.
17. It is recommended to perform document review and maintain proper control of documents.

18. Training is an essential element of the laboratory success. It is suggested to the labs to invest in training their personnel to Standard elements, in depth, choosing reputable training sources.
19. When the laboratory is a part of bigger organization, there should be clear demarcation between the divisions and firewalls in place.
20. Please note that ISO 9001 is not equivalent standard to ISO/IEC 17025. ISO 9001 fulfills some of the management requirements of the ISO/IEC 17025 standard, which leaves the technical requirements to be addressed in order to complete the accreditation to ISO 17025. ISO 9001 does NOT address any of the technical requirements of a Laboratory's management system.
21. Laboratories have to be careful to NOT use ISO 9001 certification logo on the test reports they are issuing. This is a requirement of the standard ISO/IEC 17021-1 [8], clause 8.3.2, that clearly states that "a certification body (means the ISO 9001 registry) shall not permit its marks (means ISO 9001 logo) to be applied to laboratory test, calibration or inspection reports or certificates".

5 Conclusions

Through a careful reading of the performed analysis it is evident that the majority of the identified non-conformities are related to the following ISO/IEC 17025 requirements:

- Equipment calibration and maintenance
- Estimation of measurement uncertainty
- Test methods
- Measurement traceability
- Internal audits
- Lack of root cause analysis in coming up with corrective actions
- Management review
- Lack of addressing key requirements and topics during Management Review
- Lack of commitment of top management (i.e. not participating in management reviews)
- No clear demarcation between the divisions, when part of a bigger organization
- Uncertainty of Measurement not calculated
- Laboratory key personnel not in position to explain Uncertainty of Measurement mechanics
- Lack of a plan for assuring the quality of test results
- No evidence of assuring the quality of all test results
- No evidence of intermediate controls
- Monitoring and Updating employees on new testing methodologies
- Evaluating employee competencies on periodic basis
- Lack of efforts by the laboratory to take part in proficiency testing, inter-lab comparisons
- Test reports/calibration certificates partially meeting the reporting requirements

Laboratories need to pay attention to implement effectively the corrective actions (with non-conformities) besides the policies and procedures in place, to avoid systemic failures that can result in an ineffective management system leading to drastic performances. Random failures can be controlled to only certain extent.

All systemic failures need to be controlled in an efficient way. That can be accomplished with total commitment of management (from top to bottom, including all the employees) by adhering to perform to meet the policies, procedures and requirements of the Standard and the Laboratory's Management System. Some opportunities for improvement are also suggested at the end this paper and the laboratory can work on creative processes to make sure they have an effective management system in place.

Finally, in order to facilitate the reader, at the Annex at end of this paper a table is included, presenting the corresponding clauses of 2005 and 2017 versions of the standard ISO/IEC 17025.

Appendix

See Appendix Table 1.

Table 1. Relating the clauses of ISO/IEC 17025:2005 to ISO/IEC 17025:2017

Clause version 2005	Title	Clause version 2017	Title
4.3	Document control	8.3	Control of management systems documentation (Option A)
4.6	Purchasing services and supplies	6.6	Externally provided products and services
4.13	Control of records	8.4	Control of records (Option A)
4.14	Internal audits	8.8	Internal audits (Option A)
5.2	Personnel	6.2	Personnel
5.4	Test and calibration methods	7.2	Selection, verification and validation of methods
5.5	Equipment	6.4	Equipment
5.6	Measurement traceability	6.5	Measurement traceability
5.9	Assuring the quality of test and calibration results	7.7	Ensuring the validity of results
5.10	Reporting the results	7.8	Reporting the results
4.1	Organization	5.0	Structural requirements
4.2	Management System	8.2	Management System documentation (Option A)
4.4	Review of requests, tenders and contracts	7.1	Review of requests, tenders and contracts

(*continued*)

Table 1. (*continued*)

Clause version 2005	Title	Clause version 2017	Title
4.5	Subcontracting of tests and calibrations	6.6	Externally provided products and services
4.7	Service to the customer	8.6	Improvement (Option A)
4.8	Complaints	7.9	Complaints
4.9	Control of non-conforming testing and/or calibration work	7.10	Nonconforming work
4.10	Improvement	8.6	Improvement (Option A)
4.11	Corrective action	8.7	Corrective actions (Option A)
4.12	Preventive action	–	–
4.15	Management review	8.9	Management review (Option A)
5.1	Personnel	6.2	Personnel
5.3	Accommodation and environmental conditions	6.3	Facility and environmental conditions
5.7	Sampling	7.3	Sampling
5.8	Handling of test and calibration items	7.4	Handling of test and calibration items

References

1. The new ISO/IEC 17025:2017, Dr. George Anastasopoulos, CAL LAB the International Journal of Metrology, pp. 30–35, July 2017
2. ISO/IEC 17025:2005, General requirements for the competence of testing and calibration laboratories, ISO (2005)
3. ISO/IEC 17025:2017, General requirements for the competence of testing and calibration laboratories, ISO (2017)
4. ISO/IEC 17025 moves to final stage of revision, Sandrine Tranchard, September 2017. ISO site: https://bit.ly/2ACimjY
5. Facts and Figures, Table: Total Number of Accredited Conformity Assessment Bodies (CABs) (2017). ILAC site: https://bit.ly/2ANHdBt
6. ILAC MRA and Signatories (2019). ILAC site: https://ibit.ly/2ACjcx8
7. IAS Accreditation Criteria for Testing Laboratories AC89, September 2018. www.iasonline.org/wp-content/uploads/2018/10/89-Sep-2018.pdf
8. ISO/IEC 17021-1: Conformity assessment—Requirements for bodies providing audit and certification of management systems. ISO (2015)

Novel Approach for Designing the Deployment of Urban Wifi Based on an Urban 3D Model and a Web Tool

Iñaki Prieto[1]([⊠]), Jose Luis Izkara[1], Sara Diez[2], and Mauri Benedito[3]

[1] Sustainable Construction Division,
Tecnalia Research & Innovation, Derio, Spain
{inaki.prieto,joseluis.izkara}@tecnalia.com
[2] R&D Department, Ambar Telecomunicaciones, Santander, Spain
sdiez@ambar.es
[3] Estudios GIS, Vitoria-Gasteiz, Spain
mbenedito@estudiosgis.com

Abstract. The usage of WIFI and in particular the urban WIFI has exponentially grown in the last years. However, the deployment of an urban WIFI system is a tedious labour that needs a lot of field work in order to deploy and later validation of the system. In this paper a novel approach for deploying urban WIFI is presented which is based on an urban 3D model and by mean of a web tool. The main objective of this tool is the development of a system that allows to obtain, from a 3D urban model, a catalogue of specifications and predefined rules; a proposal of solutions that indicate the ideal positions and the types of WIFI devices needed to increase the coverage offered at the minimum cost. In order to do that, the tool uses a web 2D and 3D viewer of coverage distribution maps and as it is possible to configure the catalogue, the predefined rules and all steps in the tool, as a result the needed access points and antennas are minimized.

1 Introduction

For the first time in Europe, the 85% of the users had an internet connection in 2016, according to the "Internet usage in Europe - Statistics & Facts". Therefore, users increasingly demand internet connection in different places, such as hotels or means of transport, etc. Furthermore, the internet was used daily by 71% of EU citizen, which shows that the massive use of the internet is evident (The Statistics Portal, n.d.).

The purpose of wireless technology, as its name suggests, is the interconnection of users and devices to applications and/or network services without wiring. In recent years, wireless technologies are acquiring great relevance for citizens, companies and public corporations, and in general for society, mainly due to the great flexibility in mobility in the access they provide to users. The advantages versus the traditional wired networks are: mobility, simplicity and speed of installation, flexibility of installation, cost reduction, etc. (Pahlavan and Krishnamurthy 2011).

WIFI wireless transmission technology has grown incredibly in recent years, and currently it is possible to perform many tasks that were previously impossible if the

L. Mohammad and R. Abd El-Hakim (Eds.): GeoMEast 2019, SUCI, pp. 35–48, 2020.
https://doi.org/10.1007/978-3-030-34187-9_3

people were not at home or in the office. It is just enough with a laptop or Smartphone and a point of connection to the Internet. Such is the usage of this technological combo in our lives, that it is estimated that more than half of Internet users connect to the network of networks through this method (Song and Issac 2014).

Barcelona (Spain) is one of the cities that strongly supports this type of services. Currently they have deployed more than 700 WIFI access points. In addition, the city of Barcelona will invest 3.6 million euros to implement the WIFI service in public transport and will expand it also to parks and gardens. Currently they usage of free WIFI service is near to 100.000 users per month in a city with a population of 1.6 million, which clearly shows the good acceptance of these service by citizens (Ajuntament de Barcelona, n.d.).

The implementation of an urban WIFI system currently includes a list of tedious aspects, such as: (a) network specifications, (b) dimensioning and determination of the equipment, (c) radio planning, (d) calculation of the level of radioelectric emissions, (e) deployment, (f) certification and commissioning, (g) network management and service provision, (h) recognition of the coverage area, (i) location of the wireless access point site, (j) connection to the wired network and (k) access securing (Henry and Luo 2002).

The following additional considerations will be taken also into consideration, such as consideration of public service and what it implies in terms of administrative authorizations and legality, quality of service and implications in terms of sizing of network resources, quality of service, management and security.

Security is also one of the main characteristics to consider when deploying wireless services. In addition, it is necessary to achieve a series of challenges in the deployment of WIFI networks such as efficient management of communication networks, security, optimal coverage, convergence and deployment of services over networks and speed (Potter 2006).

But on the other hand, the risks that are taken when using urban WIFI in public places seriously compromise people security. That is why it is necessary to be aware of the situation and take a series of precautions in order to make surfing on the Internet a pleasant experience (Afanasyev et al. 2010).

The use of a public network can bring with it a series of disadvantages that can be very harmful such as the possibility of being victims of data theft, since the traffic in it can be easily monitored and all our information, such as passwords for access to email accounts, credit card data and other sensitive information, collected.

In this paper a novel approach for deploying urban WIFI is presented which is based on an urban 3D model and by mean of a web tool. Indoor WIFI planning is a topic that has been deeply analyzed and there are multiple tools are available. However, outdoor planning is still an open issue where more research is needed.

The rest of the article is structured as follows. First, the current way of deploying an urban WIFI is explained. Then the new proposed novel approach is described in Sect. 3. In Sect. 4 the validation of the proposed workflow is done through a case study in Donostia/San Sebastián (Spain). Finally, the main conclusions obtained from the work described in this article are presented.

2 Current Workflow

Traditionally, the implementation of urban WIFI has focused on providing coverage to areas unique in which the influx of a small number of customers was foreseen. With the new technologies such as M2M or IoT, this type of deployment resembles that of urban mobile networks differing in that there will be fewer radiant equipment and with a separation less.

To provide these services it is necessary to carry out a preliminary study and analysis of how to offer these services with a high quality (bandwidth and range) by minimizing the number of repeaters placed in the city.

When deploying an urban WIFI both economic and technical aspects are analysed. In order to do that, the following aspects are analysed:

- **Analysis of the situation.** The main objective is to use the minimum number of AP. For that, the radioelectric spectrum can be used efficiently by reusing frequencies.
- **Radioelectric propagation and coverage.** Through the radiation diagrams of the available antennas, a propagation analysis is carried out with the areas covered by those antennas.
- **Planning in interiors and exteriors.** Coverage and capacity are related in such a way that as the distance between transmitter and receiver increases, the SNR (signal-to-noise ratio) worsens, being necessary to retransmit erroneous packets, or adopt a less efficient but more robust modulation. This aspect will be much more critical for the case of exteriors than in coverages inside buildings. In the exterior areas, compliance with local regulatory regulations, both in terms of construction and the environmental or legal impact of each area, will be essential.
- **Radiation of the antennas.** When carrying out a coverage study, the radiation diagrams of the antennas will be understood fundamentally, in order to properly understand the area of coverage provided by them. To better understand the behaviour of the antennas, an image can be shown in which it is possible to interpret the azimuth (horizontal), the elevation (vertical) and its three-dimensional composition.
- **Position of the antennas.** The strategy of deploying the antennas and the position of them must be done according to the position of the potential clients. Although in the zones of the great lateral distances antennas with great lateral advance will be used. The areas are focused on areas of reduced interest, the use of antennas with transversal orientation is used.
- **Verification with client devices.** As it has been seen, radiation in the AP direction towards the client is a good measure to be able to establish the coverage area of an antenna. One question to ask is: To what extent can we consider the operational limit of an AP coverage? The answer to this question is answered indirectly since the limiting factor will be the client device. In general, the emission capacity of an antenna is an important energy consumer, so in general portable devices will not have large emission capabilities.

Because of its increasing adoption, WIFI technology will continue to become faster, more secure, more reliable and more resourceful. These advances in turn drive new developments to incorporate this technology in the new processes to provide different services to users on an ongoing basis.

Starting from infrastructure of conventional communications systems based on passive materials, such as fiber optic cabling or copper wiring, new WIFI wireless technologies are generated with bandwidth and quality of service solutions that offer customers better services, greater availability, greater security and redundancy.

But the use of access systems has grown exponentially and is changing significantly as it is used to access the point-in-time applications of a few users and has now been used to concentrate a high density of data for the use of all types of applications.

In addition to security, the improvement of network management has become an important point. It is important to plan the networks towards the convergence of mobile applications and the density of users that it will have, therefore the choice of architecture in these deployments is key. The right choice of wireless technology at this time is the key to a network prepared for the next generation of SMART services.

3 Proposed Workflow

In this section a novel approach for deploying urban WIFI based on an urban 3D model and a web tool is explained. In order to do that, a list of actions is presented:

- **Complete the Catalogue**
  This module allows completing the system with the different elements that can be installed in an urban WIFI installation. Elements such as access points, antennas and the mounting kit can be completed in this step. For each of them, a list of parameters has to be fulfilled. In the Table 1 the defined parameters are presented. The completion of the catalogue data can be done in any moment, is independent of the projects. And in the same way, it must be maintained up to date.
- **Create Urban 3D Model**
  The urban 3D model is the urban information server, which contains geometric information such as semantics at different levels of detail or LoD (Level of Detail). Each user or application will work with the level of detail that needs. For this application, the urban 3D model is based on existing standards.
  The standard CityGML, which is defined by the OGC, is used to represent the urban 3D model. CityGML defines 5 levels of detail, from 0 to 4. Level 0 represents the 2D terrain. Levels 1 to 3 define the building in greater detail but focusing on the exterior and LoD 4 already contemplates interior elements of the building (columns, stairs, doors …).
  Then, using the LIDAR and DTM data, the real height of the buildings is obtained. In this way it is possible to generate the buildings in 3D with its real height and positioning it correctly, both in position and in altitude (on the digital terrain model). For this application building are generated in LoD2 (see Fig. 1).

Table 1. Catalogue elements and their parameters

Access point (AP)	Antennas	Mounting kit
Model	Model	Model
Manufacturer	Manufacturer	Manufacturer
Band	Band	Compatible AP
Number of antennas	Gain	Long
WIFI type	Radiation	Width
Installation	Assembly	Price
Integrated antennas	Price	
Price		

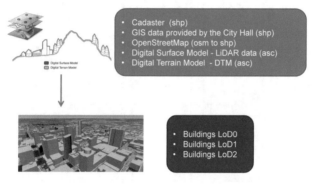

Fig. 1. Urban 3D model modelling

The urban furniture such as street lighting of the area of interest are also added since they can be optimal locations to place WIFI repeaters. Trees are also modelled because they attenuate the WIFI signal, and they need to be taken into account.

- **Identification of Optimal Locations**

  The identification of optimal locations is the main process to be carried out by the system. The objective is to identify the locations of the access points and antennas in order optimize them, but always providing coverage to the entire selected area. The Fig. 2 shows the followed workflow in order to identify the optimal locations.

When a city wants to cover a specific area with WIFI, the first step is to get in touch with a WIFI installer. The WIFI installer will *create/edit the project* in the system and will notify the GIS technician to *generate the urban 3D model* for that project.

GIS technician will then *generate the urban 3D model* and will *store* it in *CityGML* format (.gml) and export the geometry to *KML* format (for visualization purposes).

Meanwhile the WIFI installer will *create/edit a project* and its predefined rules or conditions and develop an urban 3D model of the area of interest. In the next step, a list of possible point of accesses and antennas will be shown, just the ones that apply the previously predefined rules, which are the *filtered elements from the catalogue*. From that list the WIFI installer must *select only one point of access, one antenna and one mounting kit.*

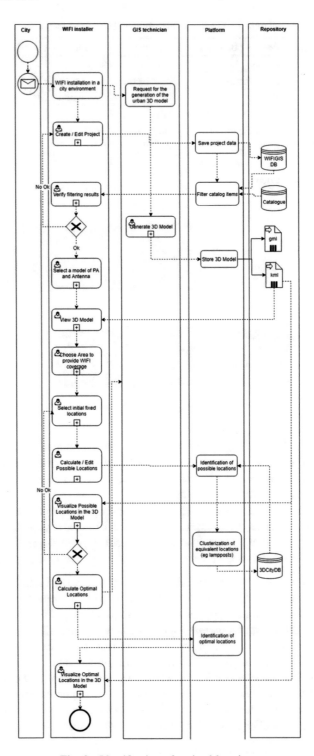

Fig. 2. Identification of optimal locations

In the next step, the WIFI installer can *visualize the generated urban 3D model* that includes buildings, street lighting and trees. Then, it must *select the area where the WIFI needs to be deployed* according to the specifications given by the city.

In the next step, which is optional, a list of *fixed positions can be defined*. Which means that, for example, if we know that in the case study is available a public building, in which it will be easy to deploy an access point, it can be defined as a fixed access point position. In this way, this access point will appear mandatorily in the final result. An advantage of this step is that, in this way, the needed simulation cycles are minimized.

In the *calculate possible locations* two different this are done. In one hand, as a constraint all the building which are public building are a possible location, because as they are public building, the needed procedures are easier than in private buildings.

In the other hand, the street lighting is a possible location for WIFI access point. However, as it is very possible that in the selected case study area a big amount of them, a clustering process is performed (see Fig. 3). In this process, which is performed in the platform, the street lighting location that can include, by distance, most of the others are selected. The number of desired clusters is specified by the WIFI installer.

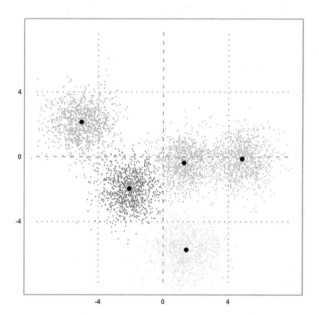

Fig. 3. Street lighting clusterization by K-means (https://rosettacode.org/wiki/K-means%2B%2B_clustering)

The next step *calculate/edit optimal locations* is iterative. In each calculation a unique new location is generated by the system. In this step all the calculations are performed in the platform, which will access to the urban 3D model information. In this way the WIFI installer can follow the simulation result and modify or delete the calculated locations if desired, until the desired result is achieved.

The following formulas are used, first in order to calculate the attenuation and then in order to calculate the received power in a certain point.

$$Lfs = 20 * log10(d) + 20 * log10(f) - 187.5 \tag{1}$$

$$Lfs = attenuation$$
$$d = distance$$
$$f = frequency\,(MHz)$$

$$Ptx + Gtx - Lfs + Grx = Prx \tag{2}$$

$$Prx = received\,power\,(dBm) \tag{3}$$

$$Ptx = transmitter\,output\,power\,(dBm)$$
$$Gtx = transmitter\,antenna\,gain\,(dBi$$
$$Lfs = free\,space\,loss\,or\,path\,loss\,(dB)$$
$$Grx = receiver\,antenna\,gain\,(dBi)$$

In addition to these formulas, as the buildings in the area can have different construction materials, different attenuation values are used. Furthermore, in case trees are in the case study, an attenuation value is also applied. In the Table 2 those attenuation values are presented.

Table 2. Attenuation values for different materials and objects (http://www.ubeeinteractive. com/support/consumers/faq/signal-attenuation-receive-signal-strength-indicator-rssi-4)

Material	Attenuation (dB)	
	2.4 GHz	5 GHz
Interior drywall	3–4	3–5
Cubicle wall	2–5	4–9
Wood door (hollow-solid)	3–4	6–7
Brick/Concrete wall	6–18	10–30
Glass/Window (not tinted)	2–3	6–8
Double-pane coated glass	13	20
Bullet-proof glass	10	20
Steel/Fire exit door	13–19	25–32
Free space	.24/ft.	.5/ft.
Tree (est.)	.15/ft.	.3/ft.

During possible and optimal location calculations, the results can be visualized in 2D and 3D in the web 3D viewer

4 Case Study

For the validation of the explained approach it has been tested in a case study in Donostia/San Sebastián (Spain).

4.1 Web Tool Architecture

The web tool architecture is split in three layers. In the data layers, two different databases have been created. The first one related with the projects data and the second one with the urban 3D model. On top of the urban 3D model a Web Feature Service have been deployed in order to access to the data contained on it in a standard way. Both databases are stored in a PostgreSQL + PostGIS database.

In the business layer the processes that need computation are developed, such as the street lighting clusterization process or optimal locations simulation. Those processes have been developed in Java.

Finally, in the presentation layer the urban WIFI deployment web tool have been developed in HTML 5 + Javascript. Two different technologies have been used for geospatial data visualization, such as Leaflet[1] for 2D data and CesiumJS[2] for 3D data.

4.2 Urban 3D model

The generation of the 3D urban model is done using different data sources as can be seen in Table 3.

Table 3. Data sources for urban 3D model generation

Element	Data source	Format	Number of elements
LIDAR	GeoEuskadi	.ASC	4
Buildings	Cadaster	SHP	506
Street lighting	City hall	DXF	1080
Trees	City hall	DXF	1208

In the Fig. 4 the resultant urban 3D model of Donostia/San Sebastián is shown.

4.3 Tool for Designing Urban WIFI Based on a Urban 3D Model

It is the responsible of calculating what types of WIFI devices should be used and the exact placement of them. The platform has as input data the urban 3D model and the catalogue. This tool is divided into the following functionalities:

---

[1] https://leafletjs.com/.

[2] https://cesiumjs.org/.

Fig. 4. Urban 3D model of Donostia/San Sebastián

- **Predefined Rules**
  This module allows to filter the elements of the catalogue using the previously defined project rules. In this way, the catalogue elements that comply with the rules defined by the user are displayed. In addition, other rules such as, maximum budget, needed coverage quality, etc.
- **Select Access Point, Antenna and Mounting Kit**
  This module allows selecting a single combination of access point, antenna and mounting kit. This combination is the one that is used later when calculating the solution.
- **Choose Area to Provide WIFI Coverage**
  This module allows selecting the area of interest in which WIFI coverage needs to be provided. This selection is made on a 2D map. In addition, the separation in meters of the coverage grid to be calculated is selected. That is, if the coverage for each of the points is calculated every 1, 2, 5, 10, 20 or 50 m. In the Fig. 5 the selection of the desired area is performed.
- **Clusterization of Street Lights**
  This module allows the clustering of the equivalent locations associated with the streetlights. If, for example, in a study area there are more than one hundred lampposts, using this module is to perform a clustering calculation to find which are the X (the number can be modified) more representative lamps. In this way, in the simulation, instead of using 500 different lamppost locations, the clustered ones are used, thus speeding up the calculation of solutions. In the Fig. 6 the clusterization of the street lights is shown. In red all the street lighting locations are presented and in green only the clusterized ones.
- **Select Initial Fixed Locations**
  Through this module the user can select fixed locations in which he necessarily wants to have an access point. In this way, that points are fixed and cannot be eliminated when the simulation is performed. This step can be done in both 2D and 3D.
- **Calculate Possible Locations**
  This module obtains the possible locations in which an access point can be placed. In case of the buildings, it must fulfil the following characteristics: to be a public

Fig. 5. Choose area to provide WIFI coverage

Fig. 6. Clusterization process result

building and to have electrical power supply. In the case of street lighting it is necessary that they have electrical power supply. In the viewer (both 2D and 3D) all the buildings and lampposts that appear in green are the possible locations. In the Fig. 7 the possible locations, both building and clusterized street lighting, are presented in green. In addition to that, two fixed positions can also be shown (red dots).

- **Calculate Optimal Locations**

  This module allows, once the clusters of the equivalent locations and selected fixed locations have been made, perform the calculation to obtain the optimal locations needed in order to have at least the minimum WIFI coverage in the whole area of

Fig. 7. Calculation of possible locations

Fig. 8. Calculation of optimal locations I

Fig. 9. Calculation of optimal locations II

interest. For this calculation also takes into account the attenuation created by both trees and buildings, thus obtaining a more realistic calculation. The result can be seen in colours on the grid of coverage intensities. This step can be done in both 2D and 3D (see Figs. 8 and 9).

- **Export Result to PDF**
  This module allows exporting the solution once it has been demonstrated that it meets all the project rules (see Fig. 10).

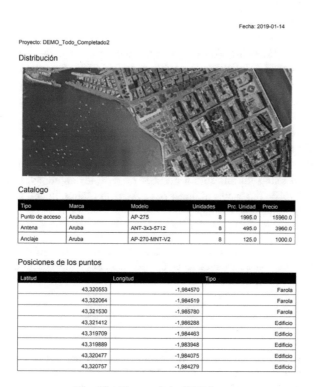

Fig. 10. The result in PDF format

5 Conclusions

The main objective of this tool is the development of a system that allows to obtain, from a 3D urban model, a catalogue of specifications and predefined rules; a proposal of solutions that indicate the ideal positions and the types of WIFI devices needed in order to increase the coverage offered at the minimum cost.

Indoor WIFI planning is a topic that has been deeply analyzed and there are multiple tools are available. However, outdoor planning is still an open issue where more research is needed.

The main aspects of innovation of the project focus on the following areas:

- **2D and 3D web viewer of coverage distribution maps.** The viewer allows to present the result of the generation of the 3D urban model. It also presents the optimal locations in order to place the WIFI devices. It also presents the result of each solution indicating the exact position of each WIFI device and present the coverage distribution map with different colours.
- **Return of investment calculation:** The existing urban WIFI simulation tools do not allow to create complete budgets. They only show the budget of the material to be used, without taking other materials or necessary works into account. Besides this, they do not allow calculating the maintenance costs and energy costs that the facilities have; what is of vital importance.
- **Urban WIFI Platform:** The urban WIFI platform have the business logic that allow the proposal of solutions on what types of WIFI devices to place and their positions. The urban WIFI platform also presents complete budgets with installation, maintenance and energy costs and the possibility of comparisons with other solutions.
- **Configurable result.** Possibility of easily adapting the configuration to the regulations and needs of different cities with different legislations.
- **Minimization of repeater antennas.** The main objective is to cover the largest coverage area using the minimum number of repeater antennas. For that it is necessary to identify the optimal locations, thus obtaining fewer signal losses and increasing the coverage radius.

Acknowledgments. The work of this paper has been done as part of the project WIFIGIS "Sistema de identificación de localizaciones óptimas basado en un modelo urbano 3D para el despliegue de WIFI urbano" with reference ZL-2017/00211.

References

Afanasyev, M., Chen, T., Voelker, G.M., Snoeren, A.C.: Usage patterns in an urban WiFi network. IEEE/ACM Trans. Netw. (TON) **18**(5), 1359–1372 (2010)

Ajuntament de Barcelona. (n.d.): WiFi comes to public transport - Barcelona. https://ajuntament.barcelona.cat/guardiaurbana/en/noticia/my-new-post-8985_101486

Henry, P.S., Luo, H.: WiFi: what's next? IEEE Commun. Mag. **40**(12), 66–72 (2002)

Pahlavan, K., Krishnamurthy, P.: Principles of Wireless Networks: A Unified Approach. Prentice Hall PTR, Upper Saddle River (2011)

Potter, B.: Wireless hotspots: petri dish of wireless security. Commun. ACM **49**(6), 50–56 (2006)

Song, S., Issac, B.: Analysis of Wifi and Wimax and Wireless Network Coexistence (2014) arXiv:1412.0721

The Statistics Portal. (n.d.): Internet usage in Europe - Statistics & Facts. https://www.statista.com/topics/3853/internet-usage-in-europe/

Project Application of BDS + Automatic Monitoring and Early Warning Technology for the Risk Assessment of Reconstruction Highway Cutting Slope

Qiang Yan[✉], Lijia Tian, and Lin Liu

Guangxi Communications Investment Technology Co., Ltd., Nanning, China
1157991006@qq.com

Abstract. It is practically important to monitor and warn the instability risk of high-steep slope of highway during cutting. The paper presents a case study carried out in a real slope in Liuzhou-Nanning Expressway Expansion Project. Grades of instability risks of slopes are firstly classified based on qualitative and quantitative assessment, and then different monitoring methods are adopted. Then, BDS (BeiDou Navigation Satellite System), Tilt photogrammetry, Internet of things are used to establish a 3D intelligent monitoring and warning system. The system was successfully to monitor and warn the risk during removal of support, and is now working well in service. In addition, a Leica TM50 high-precision measurement robot is used to calibrate the BDS + monitoring data. The results show that the trend obtained by the intelligent monitoring system is consistent with that measured by the robot.

1 Introduction

With the development of the Chinese economy, the existing expressway can no longer meet the demand for passage, and the expansion and transformation of the expressway is gradually increasing. The expansion of the existing expressway should dismantle and excavate the support of the existing cutting slope, which will cause the redistribution of stress. If the excavation site is not properly treated, there will be a risk of instability, and even serious damage to personnel and property. Therefore, it is necessary to classify the risk of slope excavation and adopt automatic technology to monitor the slope in real time according to the classification results.

Some experts and scholars have studied the risk classification of slope excavation and the monitoring technology of slope automation. Such as Lin Junyong used analytic hierarchy process and fuzzy comprehensive evaluation method to establish the security risk estimation model of high-steep slope (Lin 2017). Majilei and others put forward the Beidou Automatic Monitoring Technology and System Optimization Scheme for Highway Slope Field (Ma et al. 2017). Tao proposed to embed the cloud server (ECS), cloud database (RDS), cloud sites and other data security storage system in the "Newton force change monitoring and early warning system" based on the multi-source data fusion technology (Tao et al. 2017). Domestic and foreign experts have

© Springer Nature Switzerland AG 2020
L. Mohammad and R. Abd El-Hakim (Eds.): GeoMEast 2019, SUCI, pp. 49–60, 2020.
https://doi.org/10.1007/978-3-030-34187-9_4

done some relevant research and obtained application results. However, there is a lack of universal assessment methods and more intuitive and intelligent monitoring systems.

So, the paper presents a case study carried out in a real slope in CHINA Liuzhou-Nanning Expressway Extension Project. Grades of instability risks of slopes are firstly classified based on qualitative and quantitative assessment, and then different monitoring methods are adopted. Then, BDS, Tilt photogrammetry, Internet of things are used to establish a 3D intelligent monitoring and warning system.

2   Project Overview

Relying-on project is located in Liuzhou (Luzhai) to Nanning Expressway expansion project K1261+900–K1262+300 section, located on both sides of the highway, and it is the road slope. We use top-down grading excavation expansion to expand the way. The slope is divided into 3 zones, and the left slope is divided into 2 areas. The length of the slope is about 400 m, and the height is 45 to 105 m. The right side is divided into 1 zone, and this time we use the right side slope as the object of study. The right side slope is about 140 m long and 55 m high, which is the soil slope, and the bedrock is exposed locally at the bottom and top of the slope. The mulch layer is thicker, according to field investigation and survey data, and the composition of the mulch is mainly the plastic state of the horned gravel red clay, and exposed bedrock is mainly moderately weathered - strong weathered dolomitic limestone.

3   Slop Risk Classification

The risk assessment of slope expansion construction is carried out by expert investigation and evaluation method and index system method, according to the risk level of the assessment, special slope excavation scheme is formulated, and automatic monitoring and early warning of the slope during excavation and after excavation support are carried out to understand the stable state during construction and operation (Guangxi Traffic Investment Group Co., Ltd., Changsha University of Science & Technology 2016; Anon. 2015).

3.1   Expert Investigation and Evaluation Method

The method of expert investigation and evaluation is a method to evaluate and predict the high slope according to the construction scale, geological conditions, engineering characteristics, inducing factors, construction environment and data integrity based on the knowledge and experience of experts who are the objects of obtaining information. The expert should have senior technical title or above, and the person in charge of the evaluation should have more than 10 years' experience, and the member should have more than 5 years' experience in project management, as well as the work experience in high slope investigation, design and construction. The evaluation expert group has 3 people, according to the construction of high slope scale, geological conditions, inducing factors, construction environment, the data integrity of five disciplines, the four risk ratings are itemized rating score Ri, namely: rank **IV** (high risk) (4 points), grade **III** (high risk) (3 points), grade **II** (medium risk) (2 points), grade **I** (low risk) (1 point). On this basis, the expert confidence index Wi is given to the sub-assessment score, and the risk grade of the slope is finally determined according to the scoring value of each expert member Dr and the average scoring value $\overline{Dr}$.

A summary of the scores and results of each expert in the right area of the slope can be found in Table 1.

Table 1. Table of K1261+900–K1262+300 section right high slope expert survey and evaluation method

Expert subitem	Expert 1		Expert 2		Expert 3	
	Point	Confidence index	Point	Confidence index	Point	Confidence index
Scale of construction	$R_1 = 3$	$W_1 = 0.75$	$R_1 = 3$	$W_1 = 0.85$	$R_1 = 3$	$W_1 = 0.85$
Geological conditions	$R_2 = 3$	$W_2 = 0.75$	$R_2 = 3$	$W_2 = 0.85$	$R_2 = 3$	$W_2 = 0.90$
Inducing factors	$R_3 = 3$	$W_3 = 0.80$	$R_3 = 3$	$W_3 = 0.80$	$R_3 = 3$	$W_3 = 0.80$
Construction environment	$R_4 = 2$	$W_4 = 0.85$	$R_4 = 2$	$W_4 = 0.75$	$R_4 = 2$	$W_4 = 0.65$
Data integrity	$R_5 = 3$	$W_5 = 0.70$	$R_5 = 2$	$W_5 = 0.60$	$R_5 = 3$	$W_5 = 0.65$
Dr	2.78		2.65		2.83	
$\overline{Dr}$	2.75					
Risk level	Grade III (High risk)					

3.2   Index System Method

The indicator system of the overall risk assessment of the safety of the high slope of the road is divided into 5 categories: construction scale, geological conditions, induced factors, construction environment, data integrity, and the establishment of the evaluation index system is according to the specific conditions of each indicator. This assessment uses the materiality sequencing method to determine the weight coefficient. Experts select important metrics from metrics that affect stability based on a certain slope. After the sequence is arranged, the weight coefficient is used to distinguish the importance of each evaluation index. Weight coefficients can be calculated directly according to the formula (1).

$$\gamma = \frac{2n - 2m + 1}{n^2} \tag{1}$$

In the formula, the gamma-weight coefficient, n-evaluation indicator (important indicator) number of items, m-importance order number, $m \leq n$. The overall risk of road safety on high slope is determined by formulas (2 and 3):

$$F = \sum X_{ij} \tag{2}$$

$$X_{ij} = R_{ij} \times \gamma_{ij} \tag{3}$$

X_{ij} - Evaluation indicator point, $i = 1, 2, 3, 4, 5$; $j = 1, 2, \ldots, n$. n is the number of important indicators included for the corresponding category I assessment indicator.

After calculating the F value, the overall risk level of road safety is determined by comparison with Table 2.

Table 2. General risk rating standard for high slope safety

Risk level	F
Grade IV (Extreme risk)	F > 60
Grade III (High risk)	$45 < F \leq 60$
Grade II (Moderate risk)	$30 < F \leq 45$

Combined with the actual characteristics and findings of the slope, the important evaluation indicators of this slope are determined as the following 8 categories: slope height, slope rate, stratigraphic rock, construction season, impact of natural disasters, engineering measures, geological data and design documents, and the weight coefficients of these 8 categories of indicators are sorted and determined in Table 3.

Table 3. Table of K1261+900–K1262+300 section right slope importance indicators and weight coefficient

Order of importance	Evaluation indicators	Weight
1	The slope height	0.23
2	Formation lithology	0.20
3	Slope shape and rate	0.17
4	The construction season	0.14
5	Natural disasters	0.11
6	Type of engineering measures	0.08
7	Geological data	0.05
8	Design documents	0.02

Based on the evaluation index and weight coefficient determined in the table above, the evaluation score of each indicator of the slope is determined, as detailed in Table 4.

According to the slope importance index and its evaluation score as determined in the table above, the overall risk F value of the construction safety of the high slope of the road is calculated in the substitute formula 2:

$$F = \sum X_{ij} = 66.20$$

Table 4. Table of K1261+900–K1262+300 section right slope importance indicator assessment score

Class	Evaluation indicators	Grade	Basic score (R_{ij})	Weight coefficient (γ_{ij})	Evaluation score (X_{ij})
Scale of construction (X_1)	The slope height (X_{11})	H = 105 m	100.00	0.23	23.00
	Slope shape and rate (X_{12})	Road slope over natural slope ratio quasi gradient value $\Delta\alpha = 20°$	87.50	0.17	14.88
Geological conditions (X_2)	Formation lithology (X_{21})	Strongly weathered bedrock	30	0.20	6.00
Inducing factors (X_3)	The construction season (X_{31})	Local annual rainfall is more than 1000 mm.	100	0.14	14.00
	Impact of natural disasters (X_{32})	Natural disasters are rare	20	0.11	2.20
Construction environment (X_4)	Type of engineering measures (X_{41})	Retaining wall works	24	0.08	1.92
Data integrity (X_5)	Geological data (X_{51})	There is one survey section on each slope and two exploration points on each section	50	0.05	2.50
	Design documents (X_{52})	One slope, one drawing, one direction, the drawing is incomplete	85	0.02	1.70

According to Table 2, the overall risk level of the safety of the right slope of K1261 +900–K1262+300 as determined by the indicator system method is: Grade IV (Extreme Risk). Comprehensive expert assessment method and index system method, need to develop a special construction plan for the slope, and carry out automated monitoring and early warning.

4 BDS + Monitoring System and Program

4.1 BDS + Monitoring System (Ouyang et al. 2005; Zhang et al. 2011; Wu et al. 2004)

BDS satellite navigation system is China's self-built and independent operation satellite navigation system, and it is to provide global users with all-weather, all-day, high-precision positioning, navigation and timing services for the country's important space infrastructure. BDS deformation monitoring system can provide users with a world-wide continuous, real-time, high-precision positioning. It measures the surface deformation of landslides and it is characterized by: small measurement error, accurate measurement, no need to look across the observation points, convenient selection points, and it can be applied to a variety of collapse and slide different deformation stage 3D displacement monitoring, and it can be observed around the clock.

(1) **System Composition**

BDS deformation monitoring station is composed of data acquisition system, data transmission system, GNSS solution system and data service platform. Data acquisition system: Data acquisition system is composed of measuring receiver, power supply system and installation protection system. Among them, the measuring receiver uses a high-precision measurement GNSS receiving module, which integrates the self-developed signal-noise filtering procedure to ensure high-quality raw data. Data transmission system: The data transmission system adopts GPRS wireless transmission technology, and realizes the real-time outgoing and disconnecting function of big data through the unique communication protocol, which ensures the data integrity. GNSS solution system: the data solution system with independent intellectual property rights, high-precision continuous real-time solution. Data service platform: remote configuration of field equipment, debugging, troubleshooting, upgrade and other operations of all integrated and data management system, and achieve true integration and intelligence.

(2) **Technical Indicators**

Shown as Table 5: All-In-One BDS-Deformation Automatic Monitoring Station Technical Parameters.

(3) **Early Warning Mechanism**

In view of the relatively single alarm method in the current specification, the early warning mechanism is refined and multi-level early warning is carried out according to the severity of deformation. According to the measurement and control values provided by the design and related monitoring specifications, the early warning status of the monitoring points during construction and operation is divided into three levels according to the severity: yellow monitoring early warning, orange monitoring warning (early warning value in specification) and red monitoring early warning. Different levels of alert is corresponding to different levels of response.

Table 5. Table of all-in-one BDS-deformation automatic monitoring station technical parameters

Equipment name	Technical indicators	Technical parameters
BDS monitoring station	Single epoch calculation accuracy	Plane: $\pm(2.5$ mm $+ 1 \times 10 - 6D)$ Altitude: $\pm(5.0$ mm $+ 1 \times 10 - 6D)$ Single epoch calculation Initialization time: less than 10 min
	Data output frequency	Support for output frequency of 20 Hz, adjustable
	Power	Direct current input of 7–36 V
	Environment	Work temperature: -35 °C–75 °C Storage temperature: -40 °C–80 °C Waterproof and dustproof: IP65
	The external interface	Power/GNSS Antenna/GPRS Antenna/RS232/RS485/WiFi/Bluetooth, etc.

(4) **Equipment Place Selection**

The reference stations and observation stations of the BDS deformation monitoring station are laid out in a certain number of proportions, i.e. each of the 1 station corresponds to 3 stations, and the observatories are distributed around the reference station, with the reference station as the core, and through real-time data compare the surface deformation of the monitoring area. Generally one monitoring reference station can cover monitoring points within a radius of several kilometers. The exact number of layings depends on the site conditions and equipment performance.

4.2   Monitoring Scheme

This project monitoring surface displacement monitoring sensor selects a Samsung eight-frequency monitoring receiver, and rain monitoring sensor selects an integrated rain monitoring station, and lays 2 camera monitoring. At the same time use the measurement robot to lay 38 monitoring points (left and right side) for periodic monitoring and review.

4.2.1   BDS Monitoring Scheme

The whole system has a total of 1 base station and 6 monitoring stations, of which 2 are located on the right side slope. The base station is built in a nearby toll station, while the monitoring station is laid out according to the central line the slope, and the specific location is shown in Fig. 1. Under normal circumstances, BDS automated monitoring is implemented 24 h a day, and a set of data is collected daily on a fixed period of time to calculate the average, and the period is initially calculated according to a hydrologic year (365 days), whether to continue monitoring in the following year is according to the monitoring stability.

Fig. 1. BDS monitoring system monitoring point laying

4.2.2   High-Precision Robot Monitoring Scheme

Automatic observations are carried out using the Leica TM50 high-precision monitoring robot (Li 2017a, b), which is monitored once every 15 days and it is expected to be monitored 24 times a year. If unstable local slopes or existing cracks are found to suddenly increase during monitoring, the monitoring interval needs to be shortened according to the actual situation. According to the slope topography along the highway, the deformation monitoring network of this project should adopt a triangular network,

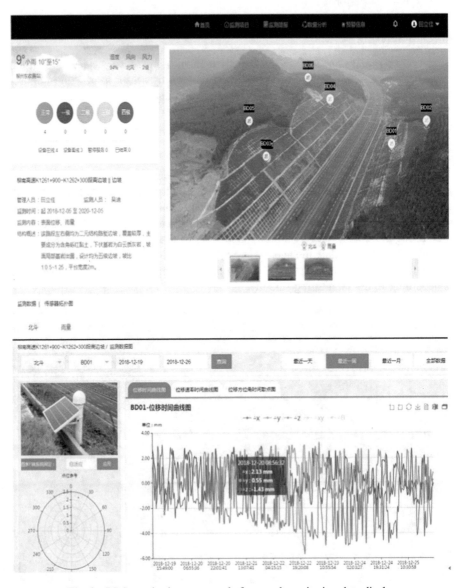

Fig. 2. BDS monitoring system platform and monitoring data display

Fig. 3. Map of the laying of robot monitoring sites

and the monitoring reference network is composed of 3 reference points and one working base point. The K1261+900–K1262+300 right-hand high-slope slope and the right-hand–turn of the slope edge to the west of 200 m belong to the outside of the slope stability range, and in this position establish the reference point TS0 (at the same time set up as a working base point), and the station near the stable position sets up a reference point, point named TS1, Along the highway to both sides of the stable position about 200 m to 500 m set a reference point, and the point name is set to TS2, TS3 (Figs. 2 and 3).

5 Monitoring Data Analysis

The validity of BDS monitoring data is verified by comparing and analyzing the data trend of the right slope BDS monitoring data and the data deformation trend of the Leica robot (Guangxi Communications Investment Technology Co. Ltd. 2019).

5.1 BDS Monitoring Data

The cumulative displacement of surface displacement monitoring pointised into cumulative horizontal displacement and cumulative settling based on the initial value

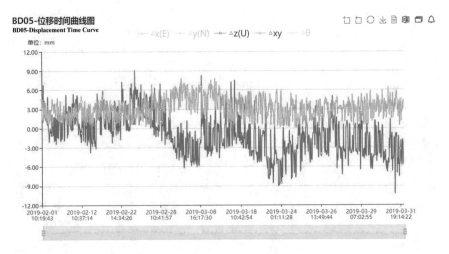

Fig. 4. BD05-displacement time curve

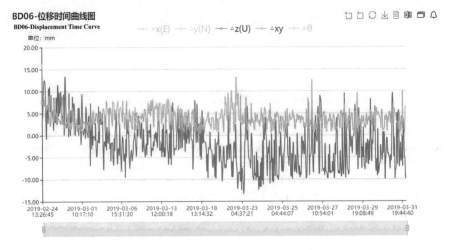

Fig. 5. BD06-displacement time curve

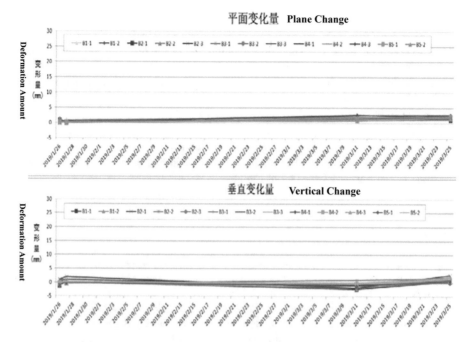

Fig. 6. Deformation amount of plane change and vertical change

of the first monitoring coordinate value. For the period from February 2019 to the end of March 2019, the cumulative displacement-time graph for the local table displacement monitoring points is shown in Figs. 4, 5 and 6. In the local time period data fluctuates greatly, mainly because the data is reduced by the poor accuracy of system solving caused by the congestion of GPRS data transmission channel. This fluctuation does not affect the determination of the deformation trend of each monitoring point.

During the monitoring period, there was no obvious trend of horizontal displacement and settlement, and the slope at the monitoring point was stable.

During the monitoring period, there was no obvious trend of horizontal displacement and settlement, and the slope at the monitoring point was stable.

5.2 Leica Monitoring Data

Using Leica periodic monitoring, the monitoring data from the end of January to the end of March were analyzed, according to the results of the monitoring data, the surface monitoring deformation rate and cumulative deformation were both small, and there was no super-warning value. The deformation trend was flat, and there was no sign of increasing.

The comprehensive comparison of the monitoring data of BDS + system and the Leica robot is consistent with the deformation trend, which shows that BDS-system can effectively play the role of automatic monitoring.

6 Conclusions

(1) The revised method of expert investigation and evaluation and the index system method are adopted to evaluate the risk of highway roadbed excavation, the evaluation results show that the slope has great excavation risk, based on this, a special monitoring program is developed.

(2) A 3D intelligent monitoring system based on BDS high precision positioning technology, Tilt photogrammetry and Internet technology is developed, and the monitoring risks shall be classified and pre-warned.

(3) The results show that the BDS + intelligent monitoring system is consistent with the change trend of Leica robot monitoring data.

References

Tao, Z., Zhang, H., Peng, Y.: Frame structure and engineering applications of multi-source system cloud service platform for landslide monitoring. J. Rock Mech. Eng. **7**, 1649–1658 (2017)

Anon.: Safety risk assessment guide for high slope construction of highway road. People's Transport Publishing Co., Ltd., Department of Safety and Quality Supervisions (2015)

Guangxi Communications Investment Technology Co. Ltd.: A periodic report on safety monitoring in K1261+900–K1262+300 of Guangxi Liunan Expressway in China, s.l.: s.n (2019)

Guangxi Traffic Investment Group Co., Ltd.: Changsha University of Science & Technology. General Risk Assessment Report on the Construction Safety of High Slope Works on Expressway Reconstruction and Expansion Project of Liuzhou (Luzhai) to Nanning Section, s.l.: s.n (2016)

Li, J.: Study on Slope Monitoring System Based on Leica GeoRobot. Chongqing Jiaotong University, Chongqing (2017a)

Li, J.: Study on the precision of measuring robot elevation by TCA 2003. J. Geomat. **04**, 92–94 (2017b)

Lin, J.: Risk Assessment of Construction Safety for Highway Deep Cutting Slope. South China University of Technology, Guangzhou (2017)

Ma, J., Lin, D., Zhang, G.: Optimization design of Beidou monitoring system for highway slope in slip fault region. Highway Traffic Technology (Applied Technology Edition), vol. 4, pp. 31–33 (2017)

Ouyang, Z., Zhang, Z., Ding, K.: Slope monitoring system of Three Gorges area based on 3S techniques and ground deformation observation. Chin. J. Rock Mech. Eng. **24**(18), 3203–3210 (2005)

Wu, Z., Deng, J., Min, H.: GIS-based management and analysis system for landslide monitoring information. Rock Soil Mech. **25**(11), 1739–1743 (2004)

Zhang, P., He, M., Tao, Z.: Modification on sliding perturbation remote monitoring system and its application effect analysis. Chin. J. Rock Mech. Eng. **30**(10), 2026–2032 (2011)

Electre Method – Decision-Making Tool for the Analysis of the Roads Technical State

Toma Mihai Gabriel and Ungureanu Roxana Daniela[✉]

Faculty of Railways Roads and Bridges, Technical University of Civil Engineering Bucharest, Bucharest, Romania
tom_gabriel91@yahoo.com, roxana.ungureanu04@yahoo.com

Abstract. Roads are terrestrial communication paths specially designed for the movement of vehicles and pedestrians and part of the national transport system. Maintaining proper road conditions is essential for traffic safety and road network availability. There are a number of factors that impact the decision-making process regarding the road structure programme management, including the technical condition of the roads, traffic safety, heavy traffic, the cost of maintenance works and the low funding allocated to this sector. Thus, an effective approach is needed to attain a balance between the program of maintenance/ repair/rehabilitation activities and the works that can be executed within the limits of the available funds, so that the investment achieves the highest profitability. This paper aims to analyze the technical state of the road network in Romania using a multi-criteria decision analysis method, namely Electre method. The method was applied on only four sectors of National Road 1 (the method requires a large number of parameters: weights and different thresholds assigned by the road network authorities) and aimed at identifying those road sections with the most severe conditions regarding the technical condition of the roads.

1 Introduction

The public road network in Romania consists in all the roads open to public traffic throughout the country. According to GO no 43/1997 regarding the road regime, from the functional and administrative territorial point of view, the roads are divided into roads of national interest, roads of county interest and roads of local interest Table 1 [1, 2].

At present, about 60% of the national road network have an outdated service life and over 5,000 km of roads are in a medium and critical condition [3].

Insufficient funds allocated to regular maintenance works/current repairs and rehabilitation are a major issue facing road managers. Lack of funds limits the number of maintenance and repair works, which leads to the increase and accentuation of degradation and ultimately to increased costs in repairing damaged road sections.

It must be also borne in mind that maintenance works have a permanent and continuous character, which makes it difficult for the road administrations to take into account the fact that, at a certain point in time, the types and quantities of works that need to be carried out are not known and they are identified throughout the year.

© Springer Nature Switzerland AG 2020
L. Mohammad and R. Abd El-Hakim (Eds.): GeoMEast 2019, SUCI, pp. 61–71, 2020.
https://doi.org/10.1007/978-3-030-34187-9_5

Table 1. The public road network in Romania, from the functional point of view

Nr.	Road category	Length - km	Road administrator
1	National roads	17654	National Company for Road Infrastructure Administration
2	Motorway	763	
3	County roads	35149	County council
4	Local roads	33296	Local council

The decision makers and road network administrators respectively often carry out rehabilitation works on the road network without taking into account the prioritization of road sections requiring maintenance works and without using a systematic procedure. These arbitrary decisions do not guarantee the efficiency of budget allocation.

Thus, road administrations need to strike a balance between the programme of maintenance/repair/rehabilitation activities and the works that may be executed within the limits of the available funds so that the investment reaches the highest return and the road network is durable.

This is possible through the adoption of a road infrastructure management system that allows optimal allocation of funds available for road maintenance and rehabilitation [4].

An efficient road infrastructure management system should answer the following questions:

- what is the technical state of the road network?
- where are maintenance/repair works required?
- which are the priority jobs?
- how much money should be allocated for the work?

The decision-making process is part of the process of solving infrastructure problems. This includes: identifying issues that can be addressed by building a new infrastructure, rehabilitating existing infrastructure or improving the management thereof. Infrastructure management involves making decisions on the maintenance, reconstruction, improvement or upgrading of the infrastructure systems.

Over time, various methods have been developed in the referenced literature to analyze and predict the durability of road networks.

Thus, the multi-criteria methods of analysis have been applied in the decision-making processes related to the maintenance and reconstruction of the transport infrastructure.

Fancello G. et al. proposes the development of a procedure to assist road network administrators in planning works to increase road safety using a multi-criteria method called "Electre I Compliance Analysis," which uses surpassing relationships based on the analysis of the concordances and of the discrepancies. The paper presents a methodology for classifying solutions by comparing them on the basis of different variables [5].

2 Electre Method – Decision-Making Tool for the Analysis of the Roads Technical Condition

Decision-making is a continuous process correlating and harmonizing the objectives with the resources, while the decision is the result of the processing of information by a person or group of people [6].

One of the methods used in making decisions is the Electre method (Elimination et Choix Traduisant to Réalité), developed by Bertrand Roy, as a tool for optimizing decision-making in certainty conditions.

By describing the use procedure designed by Bertrand Roy, the team of authors of this article analyzes and proposes the application thereof for the Romanian road network management system by creating a management work tool in the intervention strategy to make roads in exploitation durable.

Generally, from a procedural point of view, the Electre method is used in situations where there are several solutions V_i ; $i = \overline{1, m}$ possible to achieve a certain objective, while the evaluation considers several criteria C_j ; $j = \overline{1, n}$, and compares the solutions two by two [6].

2.1 Brief Theoretical Presentation of the Method

The Electre method is a ranking and choice method in the presence of multiple points of view. It allows for the variables to be sorted according to complex criteria by successive comparisons, two by two. There are several stages in the application thereof:

- establishing the decision-making solutions (V_i ; $i = \overline{1, m}$) and the resulting consequences in certain dimensions thereof (when used on roads it can refer to the characteristics of the technical condition of the roads, elements related to the traffic safety, etc.) by considering an assembly of criteria (C_j ; $j = \overline{1, n}$) that make their occurrence conditional. Each alternative is evaluated according to the established criteria, and matrix M of the qualifiers is built with the help of the awarded qualifiers:

$$M_2 = \begin{array}{c} V_1 \\ V_2 \\ . \\ . \\ . \\ V_m \end{array} \begin{pmatrix} N_{11} & N_{12} & \cdots & N_n \\ N_{21} & N_{22} & \cdots & N_{2n} \\ . & . & . & . \\ . & . & . & . \\ N_{m1} & N_{m2} & \cdots & N_{mn} \end{pmatrix} \qquad (1)$$
$$\qquad\quad C_1 \quad C_2 \quad \cdots \quad C_n$$

- for the resulting decision-making consequences, the utilities are established and then the utilities matrix;
- establishing the importance coefficient vector k_j ; $j = \overline{1, n}$ by which the decision maker adjusts the share of the criteria in the process for the making of the final decision. The sum of the coefficients of importance is equal to 1 (or 100);

- calculating the concordance matrix and the discordance matrix. The concordance matrix is square and expresses the superiority of V_i variant compared to V_j variant, the calculation of the elements being realized with the help of the reports:

$$c_{ij} = \frac{\sum\limits_{conc=1}^{m} k_j}{\sum\limits_{j=1}^{n} k_j} \qquad (2)$$

where:

$\sum\limits_{conc=1}^{m} k_j$ is the sum of the coefficients of importance corresponding to the criteria for which the rating of alternative V_i (of the homogeneous matrix M2) is greater than or equal to the rating of alternative V_j (in the homogeneous matrix M2);

$\sum\limits_{j=1}^{n} k_j$ is the sum of all the coefficients of importance, having the value of 1 or 100.

The discordance matrix is also square (m, m) and expresses the superiority of alternative V_j compared to alternative V_i, while the elements are calculated as follows:

$$d_{ij} = \frac{max(\delta d)}{h_m} \qquad (3)$$

where:

h_m is the difference between the largest and lowest grade of the homogeneous matrix M2.

$max(\delta d) = (N(V_j) - N(V_i))$, where: $N(V_j), N(V_i)$ are the grades for alternative V_j and alternative V_i, corresponding to the same criterion within the homogeneous matrix M2.

- the order of the alternatives is assessed according to the following relations:

$$c_{ij} > c_{ji}, \text{ then } V_i \geq V_j \text{ (alternative } V_i \text{ overrides alternative } V_j) \qquad (4)$$

$$d_{ij} > d_{ji}, \text{ then } V_i \geq V_j \text{ (alternative } V_i \text{ overrides alternative } V_j). \qquad (5)$$

If relation (4) is multiplied by (−1) and is combined with relation (5) then it is obtained:

$$c_{ij} - d_{ij} > c_{ji} - d_{ji}, \text{ then } V_i \geq V_j \qquad (6)$$

(alternative V_i overrides alternative V_j)

2.2   Application of the Method. Case Study

This technique has been applied by the authors of the article for the ELECTRE method calibration in the field of Road Transport Infrastructure, on 4 consecutive sectors of DN 1 in Romania.

For the multicriteria analysis, four criteria representative of the technical state of the roads (roughness index, skid resistance, bearing capacity, degradation index) were taken into account at this stage of the research. They were chosen based on the assumption that indicators must be easy to measure and must be clear to the decision-makers. The characteristics of the technical state were noted as: C1-roughness index, C2-skid resistance, C3-bearing capacity, C4-degradation index.

To exemplify the calculation, the chosen sectors had different types of degradation of the road surface. The length of each road section considered was 10 km from km 120 + 000 to km 160 + 000, and each section was numbered V1, V2, V3, V4 (as analysis solutions) to simplify the calculation.

$$M = \begin{matrix} V_1 \\ V_2 \\ V_3 \\ V_4 \end{matrix} \begin{pmatrix} 3.67 & 0.75 & 44 & 98 \\ 4.02 & 0.81 & 41 & 94 \\ 2.3 & 0.1 & 39 & 88 \\ 3.5 & 0.62 & 35 & 92 \end{pmatrix} \\ \quad\quad C_1 \quad C_2 \quad C_3 \quad C_4$$

Matrix M of the ratings was built in accordance with the provisions of the Technical Instructions for the determination of the technical condition of the modern roads CD 155-2001, for each the technical condition class, on the DN 1 sections analyzed in this Case Study, obtaining:

$$M = \begin{matrix} V_1 \\ V_2 \\ V_3 \\ V_4 \end{matrix} \begin{pmatrix} M & FB & FB & FB \\ M & FB & FB & B \\ FB & R & FB & M \\ M & B & FB & B \end{pmatrix} \quad (7) \\ \quad\quad C_1 \quad C_2 \quad C_3 \quad C_4$$

Qualitative qualifiers (FB = Very good, B = Good, M = Medium, R = Poor, FR = Very bad) for the technical criteria of the analyzed road (C1 - roughness, C2 - load bearing capacity, C3-) corresponding to M matrix were replaced by the assigned scores from 1 to 5, determined for each criterion (shares), according to CD 155-2001 and resulted in the matrix M1.

$$M = \begin{array}{c} V_1 \\ V_2 \\ V_3 \\ V_4 \end{array} \begin{pmatrix} M & FB & FB & FB \\ M & FB & FB & B \\ FB & R & FB & M \\ M & B & FB & B \end{pmatrix} \quad ==> \quad M_1 = \begin{array}{c} V_1 \\ V_2 \\ V_3 \\ V_4 \end{array} \begin{pmatrix} 3 & 5 & 5 & 5 \\ 3 & 5 & 5 & 4 \\ 5 & 2 & 5 & 3 \\ 3 & 4 & 5 & 4 \end{pmatrix} \qquad (8)$$

$$\begin{array}{cccc} C_1 & C_2 & C_3 & C_4 \end{array} \qquad\qquad \begin{array}{cccc} C_1 & C_2 & C_3 & C_4 \end{array}$$

The next analysis step takes into account the importance coefficients corresponding to each C1, C2, C3, C4 technical condition criterion assessed on each 10 km section of DN1 investigated with the ELECTRE Method.

This procedure falls to the administrator of the road as a management processing, in order to determine the optimal intervention solution, which in this Case Study is chosen as an example between V1, V2, V3, V4, as follows.

Since the characteristics of the technical condition of the roads do not influence the technical condition of the roads equally, different coefficients of importance/shares are assigned to each technical condition C1, C2, C3, C4 to take into account their individual effect. This practice is commonly found in the referenced literature, with road network administrators assigning different types of criteria to the technical condition of the roads, based on the opinions provided by the experts in the field, by the field engineers as well as by the university professors [7–13].

Thus, for case study analysis, four alternatives were established for the importance coefficient vector k_j ; $j = \overline{1,4}$ according to the intervention scenarios possible for the durability of the analyzed road. This operation is necessary to emphasize the share of the criteria in the process for the making of the final decision. The sum of the coefficients of importance must be equal to 1.

$$
\begin{aligned}
k_1 &= (0{,}25; 0{,}25; 0{,}25; 0{,}25) \text{ equal share for the criteria} \\
k_2 &= (0{,}4; 0{,}2; 0{,}2; 0{,}2) \text{ increased share for C1 roughness index} \\
k_3 &= (0{,}2; 0{,}4; 0{,}2; 0{,}2) \text{ increased share for C2 skid resistance} \qquad (9) \\
k_4 &= (0{,}2; 0{,}2; 0{,}4; 0{,}2) \text{ increased share for C3 bearing capacity} \\
k_5 &= (0{,}2; 0{,}2; 0{,}2; 0{,}4) \text{ increased share for the C4 degradation index}
\end{aligned}
$$

Matrix M1 (field matrix data) ratings are multiplied by the coefficients of importance assigned to each criterion, obtaining the homogeneous matrix M2 (matrix of technical state), the matrix of concordance (C) with the data taken from the field and the discordance matrix (D) by viewing the same field data. (The matrix of concordance is square and expresses the superiority of solution V_i compared to solution V_j).

Thus, for each scenario considered by the road manager, the parameters of matrix M2 are calculated as follows:

- for $k_1 = (0{,}25; 0{,}25; 0{,}25; 0{,}25)$:

$$M_2 = \begin{array}{c} V_1 \\ V_2 \\ V_3 \\ V_4 \end{array} \begin{pmatrix} 0,75 & 1,25 & 1,25 & 1,25 \\ 0,75 & 1,25 & 1,25 & 1 \\ 1,25 & 0,5 & 1,25 & 0,75 \\ 0,75 & 1 & 1,25 & 1 \end{pmatrix}$$
$$ \quad C_1 \quad C_2 \quad C_3 \quad C_4$$

This results in the 1.25 Very Good rating, the 1 Good rating, the 0.75 Mediocre rating and the 0.5 Injury rating in the evaluation of the technical condition indices.

On the basis of the obtained ratings, elements cij of the concordance matrix C shall be determined, as well as elements dij of the discordance matrix D:

$$C = \begin{array}{c} V_1 \\ V_2 \\ V_3 \\ V_4 \end{array} \begin{pmatrix} c_{11} & c_{12} & c_{13} & c_{14} \\ c_{21} & c_{22} & c_{23} & c_{24} \\ c_{31} & c_{32} & c_{33} & c_{34} \\ c_{41} & c_{42} & c_{43} & c_{44} \end{pmatrix} \qquad D = \begin{array}{c} V_1 \\ V_2 \\ V_3 \\ V_4 \end{array} \begin{pmatrix} d_{11} & d_{12} & d_{13} & d_{14} \\ d_{21} & d_{22} & d_{23} & d_{24} \\ d_{31} & d_{32} & d_{33} & d_{34} \\ d_{41} & d_{42} & d_{43} & d_{44} \end{pmatrix}$$
$$ C_1 \quad C_2 \quad C_3 \quad C_4 \qquad\qquad\qquad D_1 \quad D_2 \quad D_3 \quad D_4$$

concordance matrix C   discordance matrix D

- calculating the elements of the concordance matrix C_{ij} $i = \overline{1,4}$; $j = \overline{1,4}$ using the relationship:

$$c_{ij} = \frac{\sum\limits_{conc=1}^{m} k_j}{\sum\limits_{j=1}^{n} k_j}$$

where:

$\sum\limits_{conc=1}^{m} k_j$ is the sum of the coefficients of importance corresponding to the criteria for which the rating of solution V_i (of the homogeneous matrix M2) is higher than or equal to the rating of V_j alternative (in the homogeneous matrix M2);

$\sum\limits_{j=1}^{n} k_j$ is the sum of all the coefficients of importance, having the value of 1 or 100.

Using this calculation mode, the following coefficients of importance result for the matrix of concordance:

In the case of element c12 of the matrix of concordance C, the sum of the coefficients of importance is obtained after comparing the grades of solution V1 with the grades of solution V2.

$$M_2 = \begin{array}{c} V_1 \\ V_2 \\ V_3 \\ V_4 \end{array} \begin{pmatrix} 0,75 & 1,25 & 1,25 & 1,25 \\ 0,75 & 1,25 & 1,25 & 1 \\ 1,25 & 0,5 & 1,25 & 0,75 \\ 0,75 & 1 & 1,25 & 1 \end{pmatrix}$$
$$\begin{array}{cccc} C_1 & C_2 & C_3 & C_4 \end{array}$$

From the analysis of matrix M2 it can be noticed that the grades of solution V1 are higher than or equal to the grades in solution V2.

0.75 ≥ 0.75 the value of the coefficient of importance is noted, namely 0.25
1,25 ≥ 1,25, the value of the coefficient of importance is noted, namely 0.25
1,25 ≥ 1,25, the value of the coefficient of importance is noted, namely 0.25
1.25 ≥ 1 the value of the coefficient of importance is noted, namely 0.25

Thus, when adding up the coefficients of importance and applying the relation (4), we obtain $c_{12} = (0,25 + 0,25 + 0,25 + 0,25)/1 = 1$.

The other elements of the matrix are treated in the same manner, and the matrix of concordance is obtained by completing the corresponding positions Cij.

$$C = \begin{array}{c} V_1 \\ V_2 \\ V_3 \\ V_4 \end{array} \begin{pmatrix} c_{11} & c_{12} & c_{13} & c_{14} \\ c_{21} & c_{22} & c_{23} & c_{24} \\ c_{31} & c_{32} & c_{33} & c_{34} \\ c_{41} & c_{42} & c_{43} & c_{44} \end{pmatrix} \begin{array}{c} \\ \\ \end{array} \Longrightarrow \quad C = \begin{array}{c} V_1 \\ V_2 \\ V_3 \\ V_4 \end{array} \begin{pmatrix} x & 1 & 0,75 & 1 \\ 0,75 & x & 0,75 & 1 \\ 0,5 & 0,5 & x & 0,5 \\ 0,5 & 0,75 & 0,25 & x \end{pmatrix}$$
$$\begin{array}{cccc} C_1 & C_2 & C_3 & C_4 \end{array} \qquad\qquad \begin{array}{cccc} V_1 & V_2 & V_3 & V_4 \end{array}$$

The next step is the calculation of the elements of the discordance matrix using the relationship (3). The matrix of discordance is also square (m, m) and expresses the superiority of solution V_j compared to solution V_i, while the elements are calculated as follows:

$$d_{ij} = \frac{max(\delta d)}{h_m} \qquad\qquad (3)$$

where:

h_m is the difference between the higher and lowest grades of the homogeneous matrix M2.

$max(\delta d) = (N(V_j) - N(V_i))$, where: $N(V_j)$, $N(V_i)$, are the grades of solution V_j and solution V_i, corresponding to the same criterion within the homogeneous matrix M2.

In the case of element d12, the elements of matrix M2 in column C1, namely positions $N(V_j)$ = N(V2) = d21 and $N(V_i)$ = N(V1) = d11, are deducted (since the discordance matrix expresses the superiority of solution Vj versus solution Vi). The value of element d12 of the discordance matrix is obtained by dividing the difference to h_m.

$$d_{12} = |N(V_2) - N(V_1)|/h_m = (0{,}75 - 0{,}75)/0{,}75 = 0$$

The other elements of the matrix are treated in the same manner and the matrix of discordance is obtained by completing the corresponding positions Dij.

$$
D = \begin{matrix} V_1 \\ V_2 \\ V_3 \\ V_4 \end{matrix}
\begin{pmatrix}
d_{11} & d_{12} & d_{13} & d_{14} \\
d_{21} & d_{22} & d_{23} & d_{24} \\
d_{31} & d_{32} & d_{33} & d_{34} \\
d_{41} & d_{42} & d_{43} & d_{44}
\end{pmatrix}
\begin{matrix} \\[2pt] D_1 \ D_2 \ D_3 \ D_4 \end{matrix}
\ ==> \
D = \begin{matrix} V_1 \\ V_2 \\ V_3 \\ V_4 \end{matrix}
\begin{pmatrix}
x & 0 & 0{,}67 & 0 \\
0 & x & 1 & 0{,}33 \\
0 & 0 & x & 0 \\
0{,}33 & 0 & 0{,}33 & x
\end{pmatrix}
\begin{matrix} \\[2pt] V_1 \ V_2 \ V_3 \ V_4 \end{matrix}
$$

From the application of relations (4), (5) and (6) we may see that the section with the best characteristics of the technical condition is the section 3 km 140 + 000 − km 150 + 000, followed by 1 km 120 + 000 km 130 + 000 and section 2 km 130 + 000 − km 140 + 000 and 4 km 150 + 000 − km 160 + 000, the latter being foreground for the maintenance and repair works.

Similarly, the other scenarios k2, k3, k4, k5 were analysed and the road sections with a poor technical state were established.

- for $k_2 = (0{,}4; 0{,}2; 0{,}2; 0{,}2)$:

The section with the best characteristics of the technical condition is 3 km 140 + 000 − km 150 + 000, followed by 1 km 120 + 000 − km 130 + 000 and 2 km 130 + 000 − km 140 + 000 and 4 km 150 + 000 − km 160 + 000, the latter being foreground for the maintenance and repair works.

- for $k_3 = (0{,}2; 0{,}4; 0{,}2; 0{,}2)$:

The sections with the best characteristics of the technical condition are sections 1 km 120 + 000 − km 130 + 000 and 2 km 130 + 000 − km 140 + 000, followed by sections 3 km 140 + 000 − km 150 + 000 and 4 km 150 + 000 − km 160 + 000, the latter being foreground for the execution of maintenance and repair works.

- for $k_4 = (0{,}2; 0{,}2; 0{,}4; 0{,}2)$:

The section with the best characteristics of the technical condition is sections 1 km 120 + 000 − km 130 + 000, followed by sections 2, 3 and 4.

- for $k_5 = (0{,}2; 0{,}2; 0{,}2; 0{,}4)$:

The section with the best characteristics of the technical condition is section 1 km 120 + 000 − km 130 + 000, followed by sections 2, 3 and 4, the latter being foreground for the maintenance and repair works.

3 Conclusions

In order to ensure a reliable transport network, it is necessary to maintain the existing network at an acceptable level of viability, while the administrators of the road network have to strike a balance between the programme of activities involving maintenance/repair/rehabilitation works and the works that can be executed within the limit of the funds available, so that the investment achieves the highest profitability and the road network has a good durability.

A first important result of the paper is the highlight of certain global concerns about the use of multi-criteria methods for the analysis, decisions and forecast of some variables related to the technical condition of the roads.

The use of such techniques and methods has the following effects: it improves process transparency and information management; it facilitates planning and decision-making processes; it creates opportunities for outsourcing specific business processes; it reduces road use and maintenance costs.

A second result of the paper is that of describing and using a multi-criteria method - the Electre method - which assists the road manager to determine the road sections with a critical technical condition of the road network. The application of the method through the case study analyzed was carried out on only four sections of National Road 1 (the method requires a large number of parameters: different weights/shares and thresholds assigned by road authorities) and aimed at identifying those sections of the road subject to analysis with the most severe conditions in terms of technical condition. Thus, a partial classification of the road sections with the most severe conditions in terms of technical condition is obtained, namely the road section from km 130 + 000 − km 160 + 000.

The results obtained by using these procedures for assessing the technical state of a road allow the measurement of the amplitude and the severity of the problem, but this requires a consistent history. All these statistical methods and techniques are essential working tools in the field of road network management and allow the establishment of clear targets for strategic intervention actions in road transport infrastructure, further to the analysis.

References

1. Ordonanța nr. 43/1997 privind regimul drumurilor actualizat (2018). http://lege5.ro/Gratuit/geztaojw/ordonanta-nr-43-1997-privind-regimul-drumurilor
2. Lungimea căilor de transport la sfârșitul anului 2017 (2018). www.insse.ro
3. Statistici C.N.A.I.R. S.A.: Serviciul Mentenanță Drumuri Naționale și Autostrăzi
4. Gabriel, T.M.: Raport de cercetare nr. 1 – Prelucrarea statistică a datelor și interpretarea rezultatelor în vederea stabilirii caracteristicilor de viabilitate la nivelul rețelei de drumuri
5. Fancello, G., Carta, M., Fadda, P.: A decision support system for road safety analysis (2013)
6. www.sciencedirect.com
7. http://www.mpt.upt.ro/doc/curs/gp/Bazele_Managementului/Elaborarea_deciziilor_cap7.pdf
8. Shah, Y., Jain, S.S., Tiwari, D., Jain, M.K.: Development of Overall Pavement Condition Index for Urban Road Network (2013). www.sciencedirect.com

9. Ahmed, S., Vedagir, P., Krishna Rao, K.V.: Prioritization of pavement maintenance sections using objective based analytic hierarchy process. Int. J. Pavement Res. Technol. (2017). http://dx.doi.org/10.1016/j.ijprt.2017.01.001

10. Yashon Ouma, O., Opudo, J., Nyambenya, S.: Comparison of Fuzzy AHP and Fuzzy TOPSIS for Road Pavement Maintenance Prioritization: Methodological Exposition and Case Study (2015). https://www.hindawi.com/journals/ace/2015/140189/

11. Mustafa, K., Serdal, T.: Performance model for asphalt concrete pavement based on the fuzzy logic approach (2011). https://www.researchgate.net/publication/261223307_Performance_model_for_asphalt_concrete_pavement_based_on_the_fuzzy_logic_approach

12. Shah, Y.U., Jain, S.S., Parida, M.: Evaluation of prioritization methods for effective pavement maintenance of urban roads (2012). https://www.researchgate.net/publication/254255395_Evaluation_of_prioritization_methods_for_effective_pavement_maintenance_of_urban_roads

13. Juang, C.H., Amirkhanian, S.N.: Unified pavement distress index for managing flexible pavements. J. Transp. Eng. ASCE **118**(5), 686–699 (1992)

Site Characterization of Al-Burrulus Clay Formations

Ahmed M. Nasr[(⊠)], Amr F. Elhakim, and Mohamed I. Amer

Soil Mechanics and Foundation Engineering, Public Works Department,
Faculty of Engineering, Cairo University, Giza, Egypt
ahmed.shehab9292@gmail.com, aelhakim@hotmail.com

Abstract. An extensive subsurface investigation program was executed to characterize the formations in the city of Al-Burrulus site in the north coast of Egypt. The site is situated on the Mediterranean Sea coast with shallow ground water at the area. According to the subsurface ground investigations, thick layers of sands with relative densities varying from very loose to very dense sand underlain by very soft to hard clay and intermixed soils are characteristic of this area. A plethora of high quality laboratory and insitu tests were performed to identify the soil properties. The field tests include the Standard Penetration Test (SPT), Piezocone (CPTu), Seismic Piezocone (SCPT) and Downhole Test (DHT). Additionally, "undisturbed" specimens are extracted from the cohesive soils using Shelby tube samplers, while disturbed samples are obtained from cohesionless soil layers. Several laboratory tests were performed on the extracted specimens for classification purposes (e.g. grain size distribution and Atterberg limits). Tests on "undisturbed" samples are conducted to determine the strength properties (e.g. consolidated undrained triaxial test, unconsolidated undrained triaxial test, and direct shear tests). Site specific correlations are developed between the clay strength obtained from tests performed on high quality Shelby tube specimens and insitu tests. These correlations are beneficial to estimate clay strength in future projects in Al-Burrulus area based on field tests.

1 Introduction

Soil is characterized by a significant degree of variability/uncertainty in its properties. Proper selection of soil parameters is important for the analysis and design of any civil engineering project. Traditionally, boreholes are executed to extract soil samples at regular intervals with depth. The soil specimens are transported to the laboratory for classification and further characterization. Index laboratory tests are performed be to confirm visual soil classification. Soil strength parameters were determined using triaxial and direct shear tests. Typically, laboratory tests are time consuming and expensive. Noting the fast construction projects nowadays, other more economic subsurface investigation techniques are needed to provide reliable results in a shorter timeframe. On the other hand, insitu tests provide a faster and more economic means to characterize the subsurface. At Al-Burrulus site, several insitu tests, including the Standard Penetration Test (SPT), Piezocone (CPTu), Seismic Piezocone (SCPT) and Downhole Test (DHT), were executed to characterize ground conditions. These tests

© Springer Nature Switzerland AG 2020
L. Mohammad and R. Abd El-Hakim (Eds.): GeoMEast 2019, SUCI, pp. 72–91, 2020.
https://doi.org/10.1007/978-3-030-34187-9_6

are used for soil classification and to quantify some soil parameters by the use of correlations. This paper focuses on quantifying the short term shear strength of clay at Al-Burrulus area in Egypt.

2 Project Site Location/Geology

Al-Burrulus site is located in the Western Nile Delta about 55 km North East of Alexandria, on the Eastern side of the Rosetta Nile branch. The site is situated on the Mediterranean Sea coast, as shown on Fig. 1. The ground consists of low sand dunes, surface salt crusts and shallow ground water. A geological map of the area is shown in Fig. 2.

Fig. 1. The project location map (source: Google Map) Link: www.google.com/map. Access date: 3 May 2018

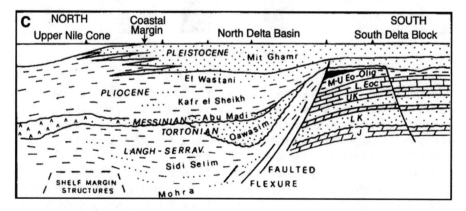

Fig. 2. Geological map of north delta basin (source: Google Images) Link: www.google.com/search/images. Access date: 13 May 2018

The ancient Nile delta ground surface now is being about 50–60 m below sea level. The sediments of the delta are marine sediments until Pleistocene time. The lower clay deposit in the northern shores of the Nile delta is extending to depth about 50 m is marine. Sand was deposited due to deposit became coarser grained, as the delta advanced. Recent alluvial and deltaic deposits are thought to be about 60–160 m thick so thick deposits of silts, clays and sands existed due to the area was infilled with marine and Nile sediments. Figure 2 shows a geological map of north delta basin.

3 Subsurface Ground Profile

An extensive subsurface investigations campaign was performed in the area of Bur-rulus site in the north coast of Egypt for the design of a major project. A total of 75 boreholes, 124 CPTu piezometers, etc... were performed. Figure 3 shows a representative a borehole log of study project zone.

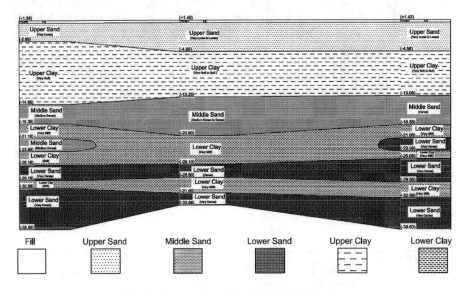

Fig. 3. Representative boreholes of study zone

According to soil investigation results in study zone, the upper 14 m soil crust is composed of very loose to loose sand extends to depth 0.6–7 m underlain by very soft to medium stiff clay with depth 0.5–8 m. While below 14 m, layers are medium dense sand extends to depth 0.6–7 m, followed by dense to very dense sand with depths 1–13 m below stiff to hard clay with depths 0.5–8 m.

In situ tests included penetration tests (SPT and CPTu) for getting soil shear strength and geophysical tests (DHT and SCPT) for obtaining soil shear modulus at small strain (G_{max}). Advantages of in situ tests are obtaining soil parameters at its natural environment (Stress state and chemical conditions) and applying different loading schemes to get soil response at various loading conditions. CPT and SCPT data were obtained within 20 m below ground level, while SPT and DH data within 40 m as shown in figures of soil subsurface investigation.

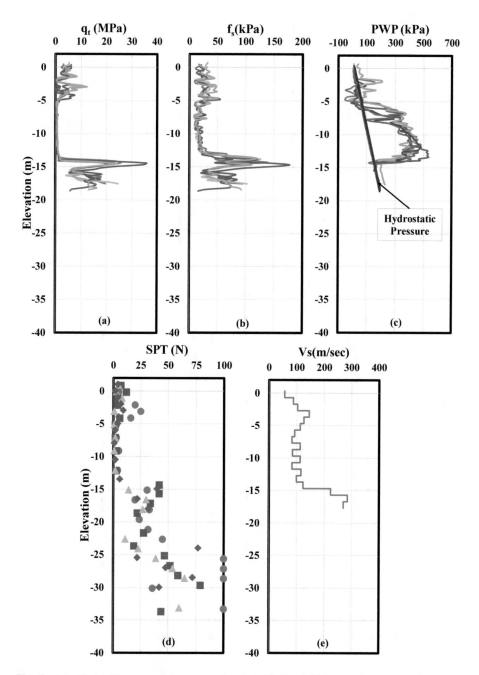

Fig. 4. (a), (b), (c) Piezocone data versus elevation (d) Standard Penetration test results versus elevation (e) Seismic Piezocone data versus elevation in Module (1) Zone (A)

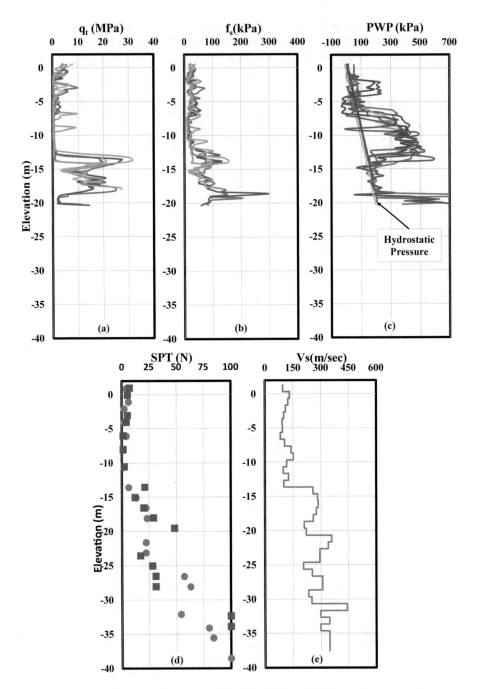

Fig. 5. (a), (b), (c) Piezocone data versus elevation (d) Standard Penetration test results versus elevation (e) Downhole data versus elevation in Module (1) Zone (B)

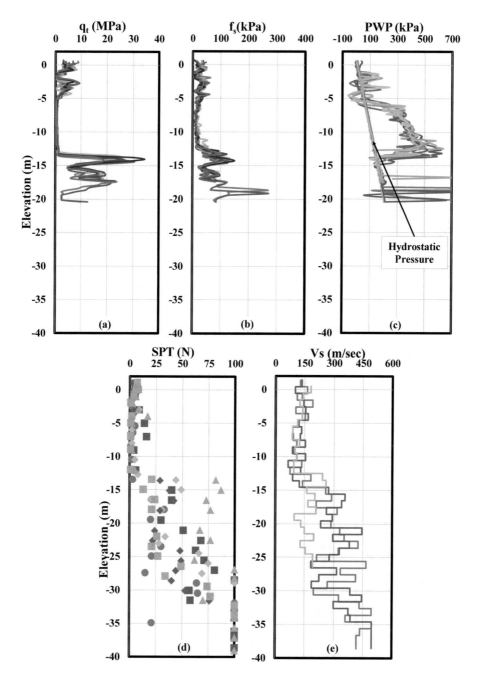

Fig. 6. (a), (b), (c) Piezocone data versus elevation (d) Standard Penetration test results versus elevation (e) Downhole and SCPTu data versus elevation in Module (1) Zone (C)

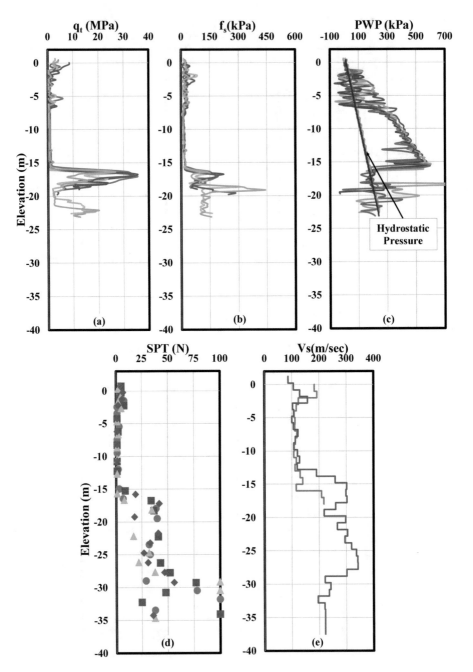

Fig. 7. (a), (b), (c) Piezocone data versus elevation (d) Standard Penetration test results versus elevation (e) Downhole and SCPTu data versus elevation in Module (4) Zone (A)

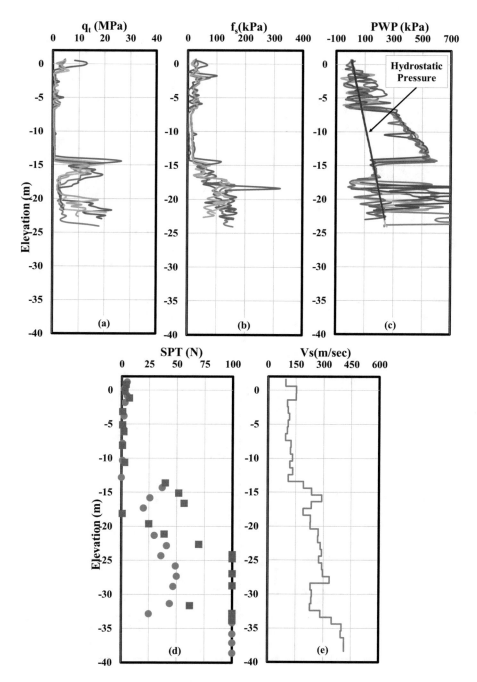

Fig. 8. (a), (b), (c) Piezocone data versus elevation (d) Standard Penetration test results versus elevation (e) Downhole data versus elevation in Module (4) Zone (B)

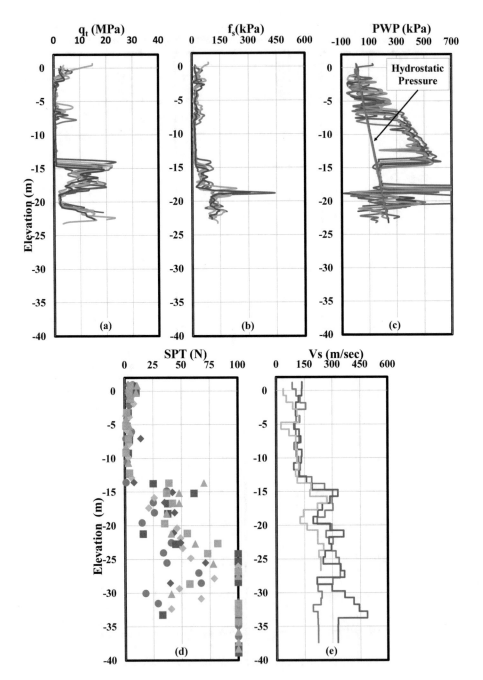

Fig. 9. (a), (b), (c) Piezocone data versus elevation (d) Standard Penetration test results versus elevation (e) Downhole and SCPTU data versus elevation in Module (4) Zone (C)

The site consists of four large modules (1, 2, 3 & 4). Each module was broken down into three zones (A, B & C) to reduce soil stratification variability. Figures 4, 5, 6, 7, 8 and 9 show data reduction of module 1 & 4 that we applied relationships of predict soil parameters.

This paper presents a comparison between the undrained shear strength parameter predicted from in situ tests and the measured parameter from laboratory tests.

4   Laboratory Tests on Clay Specimens

A plethora of laboratory tests were performed on clay specimens to determine strength parameters, as summarized below. These tests included index properties such as the liquid limit, plastic limit, shrinkage limit and water contents. The results of these tests are presented in Fig. 10. Based on the index properties, the clay is classified as CH. The consistency index varies between 0.25 and 1.65, which is indicative of very soft clay to very stiff clay as per the Egyptian Code of Practice for Soil Mechanics and Foundations (2001).

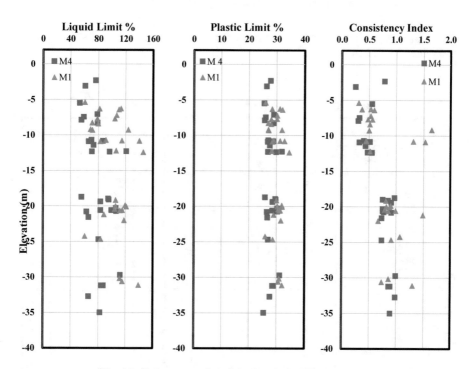

Fig. 10. Index properties with elevation of Module 1 & 4

Additionally, the stress history parameters (pre-consolidation pressure p_c and over consolidation ratio OCR) are measured from the one dimensional consolidation test performed on undisturbed samples. Alternatively, the pre-consolidation pressure (p_c') is estimated using the piezcone results as proposed by Mayne et al. (2009) and presented in Eq. (1)

$$p_c' = 0.33(q_t - \sigma_{vo})^{m'} (\frac{P_{atm}}{100})^{1-m'}$$

(1)

Where exponent (m′) is calculated the CPT soil index I_c according to Eq. (2).

$$m' = 1 - \frac{0.28}{1 + \left(\frac{I_c}{2.65}\right)^{25}}$$

(2)

Where the CPT soil index I_c is computed as per Eq. (3) (Robertson 2009)

$$I_c = \sqrt{\left[(3.47 - \log Q_{tn})^2 + (1.22 + \log F_r)^2\right]}$$

(3)

Where Q_{tn} = stress normalized cone tip resistance and F_r = normalized sleeve friction as outlined in Robertson (2009).

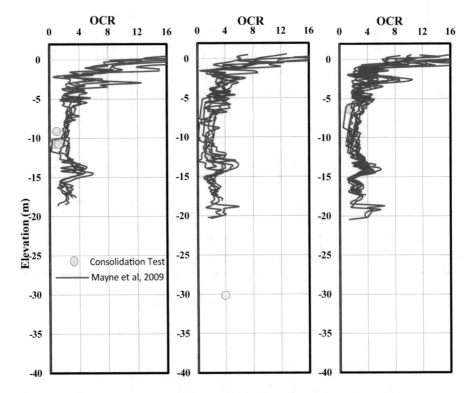

Fig. 11. Predicted versus measured OCR with elevation of Module 1

The measured and predicted values of OCR are presented in Figs. 11 and 12 for modules 1 and 4, respectively. Relatively good agreement is found between laboratory measured and piezocone evaluated values for both areas.

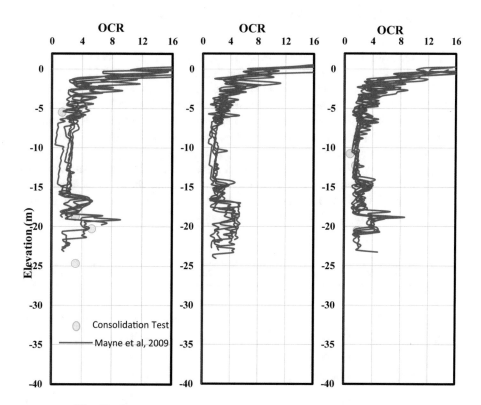

Fig. 12. Predicted versus measured OCR with elevation of Module 4

In addition, the undrained shear strength was measured by testing "undisturbed" specimens extracted using Shelby tubes. Soil strength is not a unique property but is affected many factors which include the boundary conditions, mode of loading, rate of loading, and drainage conditions. In the current study, the following tests were performed on the clay specimens to measure its undrained shear strength.

- The Unconsolidated Undrained Compression Test (UU) where drainage is not allowed during both the confining pressure and shearing stages. As no drainage is permitted, there are no changes in either the total volume or the void ratio. Although this test is not representative of the actual insitu conditions, it provides a quick and economic evaluation of strength.
- The Consolidated Undrained Compression Test (CU) where drainage is allowed during the application of confining pressure and drainage is prevented during the shearing stage. The pore water pressure may be monitored during the shearing stage. The shear strength parameter can be presented as total and effective stresses. The test may be used to measure both the undrained and drained soil strength.

– The Unconfined Compression Test (UC) is performed on cohesive soils to provide rapid approximate values of the undrained shear strength. The test is conducted on axially loaded cylindrical samples without any lateral confinement. The load applied at a high rate to prevent drainage. Usually, the measured undrained shear strength is underestimated relative to the in-situ values because of the zero confinement stresses.

The unconsolidated undrained tests are conducted at three confining stresses 100 kPa, 200 kPa, and 300 kPa that are applied to the specimen with no drainage allowed. Then the samples are loaded up to failure. Similarly, the consolidated undrained tests are performed on specimens subjected to three different confinement stresses of 100 kPa, 200 kPa and 300 kPa. The specimens are permitted to consolidate under the applied stress then loaded to failure with no drainage allowed. Pore water measurements are measured during the loading stage.

Figures 13 and 14 present the undrained shear strength values versus depth for zones 1 and 4, respectively. The laboratory test results are shown as dots while the values estimated from the field data using the correlations listed in Table 1 are presented as lines. The upper clay layer, which is approximately 14 m thick, exhibit

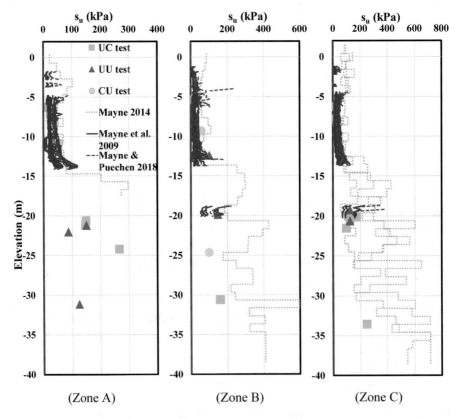

Fig. 13. Predicted versus measured undrained shear strength with Depth of Module 1

undrained shear strength that vary between 7.5 kPa and 101 kPa. The recorded undrained shear strengths are 7.5 kPa to 46 kPa, 22 kPa to 101 kPa and 38 kPa to 57 kPa for UC, UU and CU tests, respectively. The clay is described as soft to medium stiff clay according to Egyptian code of practice (2001). The predicted undrained shear strength from UC tests are estimated using Mayne et al. (2009) from the cone penetration test results and found to range between 14.5 kPa to 20.5 kPa. Similarly, the undrained shear strengths are evaluated to be 22.5 kPa to 37.38 kPa and 36 kPa to 61 kPa for UU and CU tests, respectively. The undrained shear strength from the consolidated undrained compression test is also calculated from the CPT results using the correlation proposed by Mayne and Puechen (2018). The values range between 30 kPa and 56 kPa. Finally, the undrained shear strengths are calculated using the shear wave velocity according to Mayne (2014) with values varying between 29 kPa and 109 kPa.

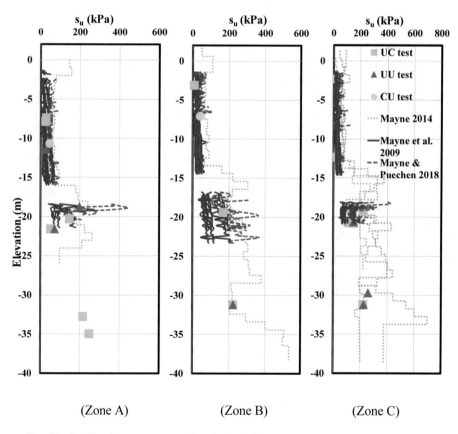

Fig. 14. Predicted versus measured undrained shear strength with Depth of Module 4

Table 1. Selected correlations for calculating undrained shear strength

Equation	Reference
$\left(\frac{S_u}{\sigma_{vo}}\right)_{UC} = 0.14\ OCR^{0.8}$ $\left(\frac{S_u}{\sigma_{vo}}\right)_{UU} = 0.185\ OCR^{0.8}$ $\left(\frac{S_u}{\sigma_{vo}}\right)_{CU} = 0.33\ OCR^{0.8}$	Mayne et al. (2009)
$s_u = \frac{q_t - \sigma_{vo}}{N_{KT}}$ $N_{KT} = 13.3$ - consolidated undrained triaxial Where $q_{net} = q_t - \sigma_{vo}$	Mayne and Puechen (2018)
$s_u(kPa) = \left(\frac{V_s}{7.93}\right)^{1.59}$	Mayne (2014)

Higher shear strength values are recorded for the lower clay layer (deeper than 14 m). The undrained shear strength measured from the unconfined compression, unconsolidated undrained and consolidated undrained are 53 kPa to 221 kPa, 73 kPa to 199 kPa and 200 kPa, respectively. As per the Egyptian Code of Practice (2001), the clay is described as stiff to hard. Using Mayne et al. (2009), the predicted undrained shear are 63 kPa to 75 kPa corresponding to unconfined compression test, 86 kPa to 109 kPa corresponding to unconsolidated undrained triaxial tests and 187 kPa for the consolidated undrained tests. While the undrained shear strengths representative of consolidated undrained triaxial loading are computed according to Mayne and Puechen (2018) with value 219 kPa. The highest estimated shear strengths are based on the shear velocity using the correlation proposed by Mayne (2014) with values of 86 kPa to 482 kPa.

All the measured and predicted strength values are presented in Fig. 15. As shown, there is a high degree of scatter in the data. The ratios of the predicted to measured undrained shear strengths are computed and presented in Table 2. The maximum average ratio is 1.8 which is highest for the shear wave based relationship proposed by Mayne (2014). For the other cases, the ratio of the predicted to measured undrained strength varies between 0.72 and 0.96. Figure 15 shows comparisons between measured and predicted undrained shear strengths for the different modes of loading. The results are bound by 2:1 line (over prediction) and 0.5:1 line (under prediction).

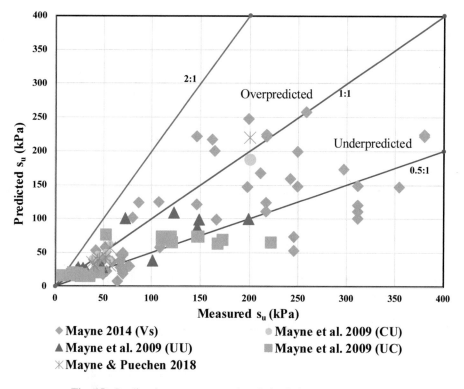

Fig. 15. Predicted versus measured undrained shear strength values

Table 2. Ratios of predicted to measured value of the undrained shear strength

	Mayne et al. (2009)			Mayne and Puechen (2018)	Mayne (2014)
	UC	UU	CU	CU	–
Minimum ratio	0.29	0.37	0.66	0.53	0.66
Maximum ratio	2.1	1.37	1.08	1.1	4.63
Mean ratio	0.72	0.81	0.96	0.86	1.8
Standard deviation	0.4	0.29	0.14	0.17	0.84

5 Site Specific Correlations

As discussed above, the soil strengths estimated using field tests show a large degree of variation compared with laboratory tests. Although some of the differences between the measured and predicted parameters may be attributed to natural soil variability, the development of site specific correlations is beneficial to improve on existing correlations. Accordingly, revised correlations are deduced to evaluate the undrained shear strength from the piezocone and seismic shear wave velocity for Al-Burrulus formations.

Consequently, the undrained shear strength obtained from the different laboratory shear tests on high quality "undisturbed" clay specimens. Figure 16 shows the correlation between the undrained shear strengths values and shear wave velocity. Correlations are developed to evaluate the undrained shear strengths for consolidated undrained triaxial, unconsolidated undrained triaxial and unconfined compression tests, as presented in Eqs. 4, 5 and 6, respectively.

$$(s_u)_{CU} = \left(\frac{V_s}{2.79}\right)^{1.12}, R^2 = 0.4 \tag{4}$$

$$(s_u)_{UU} = \left(\frac{V_s}{3.57}\right)^{1.13}, R^2 = 0.68 \tag{5}$$

$$(s_u)_{UC} = \left(\frac{V_s}{5.52}\right)^{1.27}, R^2 = 0.66 \tag{6}$$

Where undrained shear strength s_u in kPa and shear wave velocity V_s in m/s.

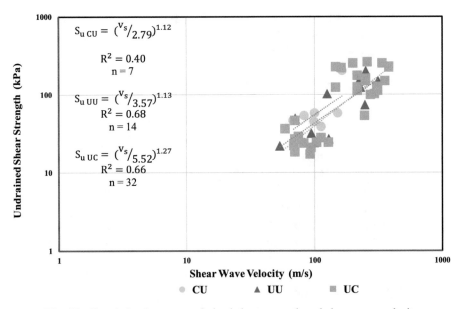

Fig. 16. Correlation between undrained shear strength and shear wave velocity

Similarly, the undrained shear strengths values are plotted versus the over consolidation ratios for the consolidated undrained triaxial, unconsolidated undrained triaxial and unconfined compression tests as shown in Fig. 17. The best fit correlations are listed in Eqs. 7 through 9 for the three loading types.

$$(s_u/\sigma'_{vo})_{CU} = 0.34(OCR)^{0.8}, R^2 = 0.9 \tag{7}$$

$$(s_u/\sigma'_{vo})_{UU} = 0.29(OCR)^{0.8}, R^2 = 0.66 \tag{8}$$

$$(s_u/\sigma'_{vo})_{UC} = 0.26(OCR)^{0.8}, R^2 = 0.46 \tag{9}$$

Finally, the undrained shear strengths are plotted versus the net cone resistance as shown in Fig. 18. Equations 10, 11 and 12 show the best fit correlations for the consolidated undrained triaxial, unconsolidated undrained triaxial and unconfined compression tests, respectively.

$$(s_u)_{CU} = \frac{q_{net}}{13.75}, R^2 = 0.93 \tag{10}$$

$$(s_u)_{UU} = \frac{q_{net}}{22.32}, R^2 = 0.67 \tag{11}$$

$$(s_u)_{UC} = \frac{q_{net}}{18.62}, R^2 = 0.75 \tag{12}$$

Where $q_{net} = q_t - \sigma_{vo}$.

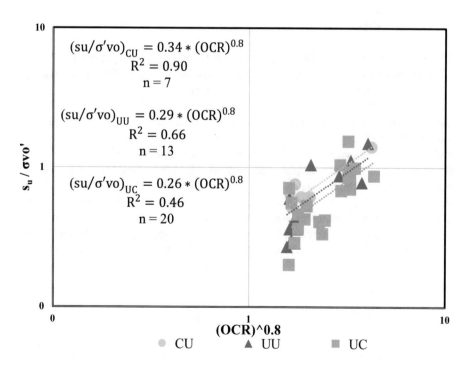

Fig. 17. Correlation between undrained shear strength and over consolidation ratio

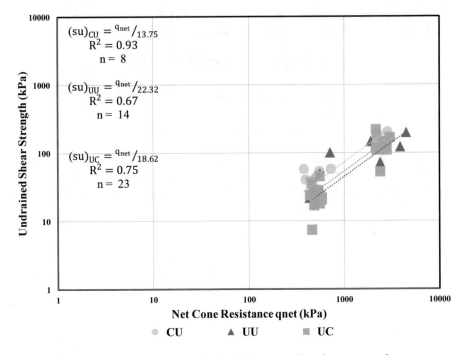

Fig. 18. Correlation between undrained shear strength and net cone resistance

6 Conclusions

This paper presents the results of an extensive subsurface investigations campaign in Al-Burrulus area in Northern Egypt. The site is situated on the Mediterranean Sea coast with shallow ground water at the area. The ground is formed of thick layers of very loose to very dense sands underlain by very soft to hard clay and intermixed soils are characteristic of this area. High quality "undisturbed" clay samples are extracted at different depths. The piezocone and shear wave velocity results are used to evaluate the undrained shear strengths for different loading types (consolidated undrained triaxial, unconsolidated undrained triaxial and unconfined compression tests) are evaluated using a number of correlations. Site specific correlations are developed to quantify clay strengths from field tests. These correlations can be used to estimate shear strengths from insitu test at Al-Burrulus area which is valuable because extracting high quality "undisturbed" soil specimens is expensive and time consuming. Thus, the use of these correlations would save cost and time on future subsurface investigation campaigns.

References

Bowles, J.E.: Analytical and Computer Methods in Foundation Engineering, pp. 44–48. McGraw-Hill Book Company, New York (1974)

Das, B.M., Ameratunga, J., Sivakugan, N.: Correlations of soil and rock properties in geotechnical engineering. Dev. Geotech. Eng. (2016). https://doi.org/10.1007/978-81-322-2629-1

Egyptian Code of Practice for Soil Mechanics, Design, and Construction of Foundations, Ministry of Housing, 283 Cairo, Egypt, vol. 3 (2001)

Elhakim, A.F.: Evaluation of soil parameters for shallow footing design from seismic piezocone test data. Ph.D. in Civil and Environmental engineering, Georgia Institute of Technology (2005)

Kulhawy, F.H., Mayne, P.W.: Manual on estimating soil properties for foundation design. Research project 1493-6 EL-6800 (1990)

Levesques, C.L., Locat, J., Leroueil, S.: Characterization of postglacial of the Saguenay Frojd, Quebec. Charact. Eng. Prop. Nat. Soils 4, 2645–2677 (2007)

Mayne, P.W., Christopher, B.R., Delong, J.: Manual on subsurface investigation. National Highway Institute Publication No. FHWA NHI-01-031 Federal Highway Administration, Washington, DC (2001)

Mayne, P.W., Coop, M.R., Springman, S., Haung, A.-B., Zornberg, J.: State-of-the-Art Paper (SOA-1): GeoMaterial behavior and testing. In: Proceedings of 17th International Conference Soil Mechanics and Geotechnical Engineering (ICSMGE), vol. 4, Alexandria, Egypt, pp. 2777–2872. Millpress/IOS Press, Rotterdam (2009)

Mayne, P.W.: Interpretation of geotechnical parameters from seismic piezocone tests. In: Proceedings of the 3rd International Symposium on Cone Penetration Testing (CPT 2014), Las Vegas (2014)

Mayne, P.W., Peuchen, J.: CPTu bearing factor Nkt for undrained strength evaluation in clays. In: Fourth International Symposium on Cone Penetration Testing (CPT 2018) Conference, At Delft (2018)

Sabatini, P.J., Bachus, R.C., Mayne, P.W., Schneider, J.A., Zettler, T.E.: Geotechnical Engineering Circular no. 5. Evaluation of Soil and Rock Properties Technical Manual, U.S. Department of Transportation, Federal Highway Administration FHWA-IF-02-034 (2002)

Robertson, P.K.: Interpretation of cone penetration tests – a unified approach. Can. Geotech. J. **49** (11), 1337–1355 (2009)

Visual Inspection Based Maintenance Strategy on Unsealed Road Network in Australia

Vasantsingh Pardeshi and Sanjay Nimbalkar[✉]

School of Civil and Environmental Engineering,
University of Technology Sydney, Ultimo, NSW 2007, Australia
VasantsinghBhimsingh.Pardeshi@student.uts.edu.au,
Sanjay.Nimbalkar@uts.edu.au

Abstract. Gravel loss on unsealed roads is financially a major setback for all road agencies. The maintenance strategy in Australia is primarily based on visual inspection only which has inherent drawbacks. It is also environmentally unsustainable. Key challenges associated with the unsealed road management are identified as (i) difficulty to forecast behaviour, (ii) significant data collection needs and (iii) vulnerability in level of service & maintenance practices. The quality of gravel material is one of the major influencing factors. Various theoretical models have developed in the past to estimate gravel loss. These studies are used to estimate road deterioration and maintenance strategy in practice. A range of approaches such as regression analysis of field measured data, system dynamic modelling approach, laboratory measurements have been employed in the past. In the state of Queensland of Australia, the long-term gravel loss monitoring is planned in the region affected by frequent flood to assess the amount of gravel loss and associated maintenance issues. Recently, a mammoth maintenance work was undertaken on unsealed road network to cater for gravel loss due to Ex Tropical Cyclone Debbie. Visual inspection can be a useful tool to provide snapshot of unseal road network at a time. Most road agencies still use this visual inspection as a maintenance strategy tool. It is recommended in this study to integrate visual inspection with localised gravel loss model and roughness data. GPS based data can be integrated by using visual inspections, roughness and the gravel model.

Keywords: Gravel loss · Unsealed road · Visual inspection · Theoretical model

1 Introduction

Australia has a staggering 574,660 km of gravel roads, comprising more than 60% of the nation's road length. Most of these unsealed roads are the responsibility of their local council, which is onerous task considering many rural and regional councils cover huge areas and have only a small population to try to meet the costs to maintain this network. Queensland State has 51,482 km of unsealed road length. It compromises 38% sealed road network and 62% unsealed road network. Scenic Rim Regional Council (SRRC) is responsible to maintain a huge network of sealed and unsealed roads. SRRC has a road

© Springer Nature Switzerland AG 2020
L. Mohammad and R. Abd El-Hakim (Eds.): GeoMEast 2019, SUCI, pp. 92–103, 2020.
https://doi.org/10.1007/978-3-030-34187-9_7

network of 1,816 kms, which compromises of 955 kms of sealed roads, 861 kms of unsealed roads and a small number of unpaved roads. SRRC has 47.42% unsealed road network and 52.58% of unsealed road network.

For about 500,000 km length of unsealed roads in Australia approximate maintenance cost per year is A 1 Billion dollars. SRRC annual maintenance budget on unsealed road is about a 2 Million dollars per year. Gravel roads are the economic backbone of Australia that depend on mining, farming and forestry. Those low traffic volume roads cannot be upgraded to sealed roads. As stated in first paragraph Australia's unsealed road network is around 60% of the whole road network and requires arbout 20% to 30% of the total road maintenance cost. It is large part of the road infrastructure cost. Societies in rural and regional Australia depend on the unsealed road network for transportation of produce and people on this road network. Major complaints across many councils pertaining to gravel roads are related to dust, potholes and either slippery road after wet weather or loose stones after long period of dry weather.

This paper provides background information relevant to unsealed roads, problems/concerns related to unsealed roads and maintenance. Emphasis is on visual inspections for maintenance management and priorities. Advantages and limitations of visual inspections and how visual inspection can be used in a better way using GPS applications to provide visual data. This paper provides a case study on unsealed road network within SRRC. SRRC unsealed road maintenance strategy is developed with visual inspection only. It also provides brief overview on how a unseal road research project was initiated due to Ex tropical Cyclone Debbie.

1.1   Challenges Associated with Maintenance and Management of Unsealed Roads

Unsealed roads considered in this research are low volume road (generally 20 to 500 vehicles per day) designed to the appropriate design standards with necessary drainage. This unsealed road has a layer of granular material which is compacted. The thickness of this compacted layer is dependent on traffic loading. It is required to maintain unsealed road regularly. It is worth mentioning that sealed roads are the ones with top layer either spray seal or asphalt which protects the underneath pavement layers from water ingress.

Dust emission and uncomfortable driving conditions are two main complaints encountered by local road authorities. Many research projects have struggled to forecast unsealed road problems. A common challenge for most of the predicting models is the intricate relationship between gravel quality (properties), maintenance practices, weather, drainage, construction practices and, geometric design. There are two tactics used for unsealed road data collection. First is to gather a substantial number of condition data regularly. Second one is road inspections by professionals to evaluate the complete condition of the unsealed network. The first tactic causes a substantial burden on road agencies, while second tactic lacks information for strategic planning. Due to neither approach is entirely reasonable, data collection methods with most road agencies varies significantly according to ad-hoc practices. Due to the flexible acceptance of specifications for unsealed roads over sealed roads, there is a large discrepancy for the levels of service on unsealed roads and the maintenance practices. There are huge variances in the geology, landscape, subgrades, environment and material between areas, which causes large variances in unsealed road performance. A reliable planning method that delivers for these variances is not possible.

'The Unsealed Roads Manual, Guidelines to Good Practice, 3rd edition March 2009' covers management procedures and practices for unsealed road network in Australian and New Zealand. This manual was developed by George Giummarra from Australian Road Research Board (ARRB). ARRB has clearly indicated that unsealed roads are presently managed with very little technical input. Due to this lack of technical input, full benefit is not realised from the available funding. The Unsealed Roads Manual provides a sound technical base for various areas relating to design, construction and maintenance of unsealed roads. Local and State Government authorities and other agencies are the major beneficiary of this manual. *Unsealed Roads Tactical Asset Management Manual (August 2015) in New Zealand* also mentions significant dependence on exhaustive data collection. The issues with gathering data related to unsealed road management systems are:

- Specific material characteristics normally do not exist for individual unsealed roads
- Unseal road condition can change significantly in short period of time, which makes existing data not relevant after it has collected
- Regular data collection is needed to for the management system adequately reliable. But regular data collection sometimes is not be cost effective and can be demanding task for road agency.

2   Unsealed Road Maintenance Types

The maintenance for unsealed roads can be categorised into two major types: periodic maintenance which is performed on yearly basis programme and monthly or predetermined frequency (2 to 6 months) routine maintenance based on cyclic programme. Periodic maintenance is long term regravelling which requires to address gravelling timing, quantity of gravel which is gravel layer thickness. Routine maintenance of unsealed road targets surface defects only for which route cause is traffic and weather conditions. Routine maintenance programme revolves around grading programme. Grading programme is governed by level of service depending on availability of maintenance funds and managing road user expectations.

In Australia different types of grading are light, medium, heavy and resheeting. A Light formation grading is to restore rideability. Where the road is formed, and loss of shape and material is minor only, a Light formation grading may be appropriate to restore shape. Medium formation grading is to restore the road surface to profile and condition. Work includes roughening of top 50 mm of roadway (by grader), clearing and grubbing to remove light vegetation and grass, recovery of suitable material from table drains (by grader), incorporation of water and compaction. No material is imported for this activity. Heavy Formation grading is to reinstatement of formation and profile of road. Work include clearing and grubbing and recovery of suitable material from table drains (by grader), tyne less than 100 mm depth, incorporation of additional gravel/material, trimming, and compaction. Resheeting is the addition of imported gravel/material to the roadway to reinstate the running surface and correct profile. Work include preparation of the formation through heavy formation grading. Supply and spreading of imported gravel/material. Imported material should be consistent with the existing gravel material.

In New Zealand grading is also divided into different types: routine grading, cut out grading, dry weather grading, wet weather grading and so on. Routine grading is to maintain the road in smooth rideable condition and removing potholes and corrugations by not changing the pavement cross section. Cut out grading is cuts in the table drain or high shoulders to drain out surface runoff from pavement. Dry weather grading is mainly during drought conditions to manage loose pavement material top of pavement surface. This loose material is deposited on the side so that it can be respread when water becomes available at later date. Wet weather grading is when the pavement has better moisture content. Ideal timing is after rain when there is optimum moisture in the pavement. Grading Cycle also depends on traffic volumes, road and weather conditions i.e. rainfall, winter, summer drought conditions, harmonization with renewal of road works e.g. the gravelling and resheeting programmes, public complaints and political pressures and level of service the client requires.

3   Unsealed Road Inspection Types

Unsealed road inspections can be classified into two major types Periodic condition surveys and routine inspections. Periodic condition surveys are GPR, Roughnesss, visual, drainage inspections. Routine inspections are monthly inspections by maintenance crew, client audits and inspections based on public complaints.

3.1   Periodic Condition Surveys

Ground Penetrating Radar (GPR) surveys can be used to record gravel thickness and other relevant data on Unsealed roads. The measuring antenna and the software used to analyse for GPR have been improved during last decade. There is a misunderstanding that GPR can provide precise gravel layer thicknesses. The GPR data analysis is time consuming and requires expert interpretation and costly. Roughometer type surveys use roughness meters and cell phone applications such as Road Roid for unsealed road data surveys. During recent years the roughometer devices which can measure roughness on an unsealed road have developed by leaps and bounds. The location is recorded using GPS (Global Positioning System) receiver and a roughness value is also recorded by a sensor attached to vehicle suspension. The results are mapped graphically due to GPS data which can be compared with previous surveys. Frequent roughness surveys are useful to compare similar problem identified. A better drainage which keeps away water from pavement plays major role in maintaining the road in good condition. Drainage Inspections are carried out by asset management team or field staff.

3.2 Routine Inspections

Field foreman perform visual inspections at predetermined frequency from monthly to quarterly of an unsealed network. This monthly inspection is for maintenance crew which carries out small repairs like potholes and associated fillings. Maintenance graders locations and activities are recorded using a netbook or tablet fitted in the grader in New Zealand. These records can be viewed on screen or exported to a GIS system for further comparison. Routine Grading can be classified one more type of inspections. The routine grader operator who has been on the network or part of network for a long period of time and has developed a better understanding of the unsealed network can be very useful to the maintenance team while carrying out grading work. This grader operator may help routine crews repairing and other maintenance activities. Most Road authorities or local councils have a customer request management (CRM) system to record and tress calls of consumers or road users. These CRM are used by road authorities to maintenance crew to inspect, analyse and determine needs for maintenance activities. Analysis of those CRM's can help for planning of routine and maintenance of unsealed road network.

4 Case Study: Unsealed Road Maintenance and Visual Condition Assessment at SRRC

SRRC area is 4,200 km^2 and 40,000 people call it a home. The region is surrounded by mountain ranges on east, south and west. On 30th and 31st March 2017 the rainfall produced by Cyclone Debbie and the cold front meeting over the SRRC ranged from 350 mm in the West of the Scenic Rim region to 800 mm in the East of the Scenic Rim region in a 24-h period. The annual average rainfall for Scenic Rim was 892 mm (Cryna weather station). The 24-h flood event was approximately equal to the annual rainfall. This high rainfall in short periods of time created ground water velocities causing sever scour damage to most gravel surfaces. Sealed roads suffered very little damage, while gravel roads and bridges received major damage. Figure 1 shows the SRRC area.

Due to large amount of gravel road involved and many roads now getting recon-structed (regravelling and resheeting) SRRC has initiated gravel road research project. Aim of this project is to enhance the existing gravel material specification, measure gravel loss on this new modified gravel material, calibrate the existing gravel loss models, refine those existing gravel loss models to better suit SRRC area. Based on those refined models the gravel road maintenance strategy will be developed.

The maintenance strategy and visual inspection is discussed briefly here. The current Maintenance Management Manual (MMM) provides guidance for program-ming and prioritisation of defects for sealed and unsealed roads. MMM provides maintenance management system with regular inspection of road network and it is anticipated that road defects will be identified and programmed for inspection. A comprehensive intervention level response time per activity listing is provided in an annexure to this manual. Where intervention levels are exceeded, these must be highlighted at Works meetings. The Works Manager is to review the list of all activities with exceeded intervention levels and reprioritise, assign more resource or employ external contractors. It is common that low risk activities are often pushed back in favour of new high-risk activities. A suggested process for this is to increase the priority level every time an activity intervention level is exceeded. In this way the activity will become high enough to receive attention.

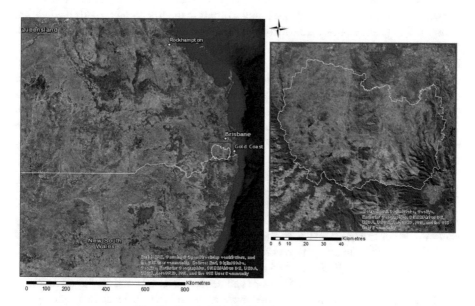

Fig. 1. Scenic Rim Regional Council map

Council engages external contractors to undertake condition assessments of sealed roads, unsealed roads and footpath. Condition assessments of unsealed roads and footpath is carried out each year. The data produced from these assessments is fundamental to assess, programme and manage road assets. Table 1 provides details of the inspection frequency based on road category. Road Category is dependent on annual traffic per day. Unsealed roads within SRRC are under category from 5a to 5d. Higher volume roads are inspected more frequently and lower volume roads are less frequently.

Table 1. Inspection frequency based on Road category

Road category	4a: rural connector	4b: rural collector	5a: rural access	5b: rural access	5c: rural access	5d: rural access	5b: rural access
Traffic volume (ADT)	1000–3000	500–1000	150–500	80–150	40–80	10–40	2–10
Inspection frequency	Monthly	Monthly	3 monthly	4 monthly	4 monthly	4 monthly	6 monthly

In addition, Council undertakes inspections on all road infrastructure on a timely manner according to the Council's road hierarchy. The higher-class roads are given higher priority in inspection whereas lower class roads receive lower priority. Defects are identified by Road Name, Chainage, Type of defect e.g. pavement failure, gravel resheeting, shoulder resheeting etc. and Maintenance Area. Condition rating is completed using an electronic platform called 'Reflect' which is asset management database software. This system uses the data entered to analyse the roads and provide a condition rating B. Safety Inspections are carried out on all roads listed on Council's Road Asset Register during the month of August on an annual basis. An additional annual Safety inspection are carried out on Roads with an estimated AADT in excess of 400 vpd during the month of February. Condition inspections are carried out annually during the months of October, November and December. This timeframe allows the information gathered to be collated and assessed for possible budget submissions. Condition Inspections may be required at irregular intervals due to unforeseen circumstances such as higher than average rainfall. In cases such as this, only the effected locations would be inspected to determine network needs.

4.1 Visual Condition Assessment

AT SRRC Visual Condition Assessment of unsealed road network is carried out by capturing video imagery and then the assessment of road segments with assessment criteria. This video imagery is recorded by asset management company. Integrated Road Survey Vehicle (IRSV) is used to capture video imagery of a road network. This imagery is recorded at every 100 m of road section intervals and in the direction of IRSV travel. The data is also provided with an accurate GIS layer representing the location of each 100 m assessment. This data is linked to a GIS road segment.

Table 2. Visual condition rating criteria

Score	Loss material	Roughness	Scouring & potholes	Corrugations	Shape
1	Nil	Very Good	Nil	Nil	Clear Crown
2	Sporadic & minor	Good	<10% & Minor	Just Perceptible	No Crown
3	Significant & substantial	Fair	<25% & Moderate	Mild < 25 mm	Mild Rutting < 25 mm
4	Extensive & substantial	Poor	<50% or Extreme	Significant < 50 mm	Deep Rutting < 75 mm
5	Extensive & extreme	Extreme	>50% & Extreme	Extreme > 50 mm	Extreme Rutting > 75 mm
9	No gravel present	Variable	N/A	N/A	Track Only

The condition assessment definitions are scored from 1 to 5, based on surface condition. Condition 9 is used to indicate an unmaintained or non-existent pavement structure which is commonly known as track. Surface condition major categories are loss of material, roughness, scouring and potholes, corrugations and shape. Based on surface condition and the score from 1 to 5, a condition rating guidance matrix is developed. Table 2 provides details of the condition rating criteria used by SRRC and external contractor to record data during visual inspection.

Condition 1, indicates a pavement in a sound state of service. Condition 2, indicates pavement in reasonable state of service. Condition 3, indicates pavement in a fair state of service. Condition 4, indicates a pavement in poor service condition. Condition 5, indicates a pavement in extremely poor service condition. These conditions rating are generic and do not consider the type of gravel. These condition definitions are widely used by many local authorities in Australia. The total extent of the specific defects or condition score for every 100 m assessment section, is calculated and reported either total extents or average conditions score to one decimal place accuracy. It is also linked to GIS road segment.

Figure 2 represents the visual average score for the entire unsealed road network. One set is for 2017 and 2 sets for 2018. There was deliberate decision to condition assess twice in a year due to rapidly changing gravel network and funding approvals. This visual data was quite useful for flood repair approvals from the funding agencies. February 2017 assessment was a month prior to Cyclone Debbie and the other two assessments are post Cyclone Debbie and flood recovery work. There are noticeable changes in network condition. During 2017 there was only 10 km of network under condition 1 and majority was under condition 2 and 3(725 km). After cyclone as the flood repair work is progressing there is noticeable increase of condition 1 (95 km). Condition 2 and 3 together are 614 km. Since 2017 to 2018 length of road under condition 1 and 2 has increased whereas length of road under condition 3 has decreased. This is because of resheeting and regravelling work completed during 2018.

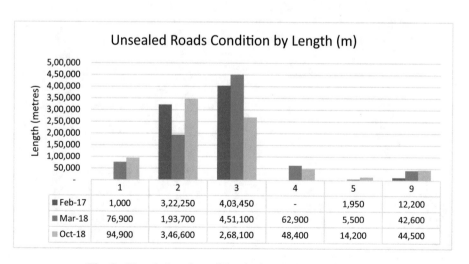

	1	2	3	4	5	9
Feb-17	1,000	3,22,250	4,03,450	-	1,950	12,200
Mar-18	76,900	1,93,700	4,51,100	62,900	5,500	42,600
Oct-18	94,900	3,46,600	2,68,100	48,400	14,200	44,500

Fig. 2. Unsealed road condition by length during 2017–2018

Figure 3 is a graphical representation of maintenance strategy based on visual condition assessment. It displays total length of road under 3 major categories: No maintenance required (combined condition rated sections of 1 and 2), Requires maintenance (condition rated sections of 3), Priority maintenance (combined condition rated sections of 4 and 5).

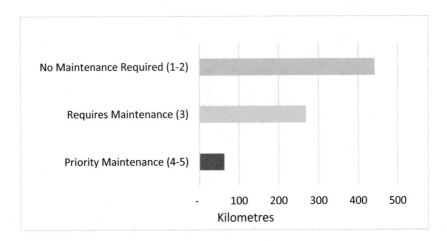

Fig. 3. Unsealed road maintenance strategy

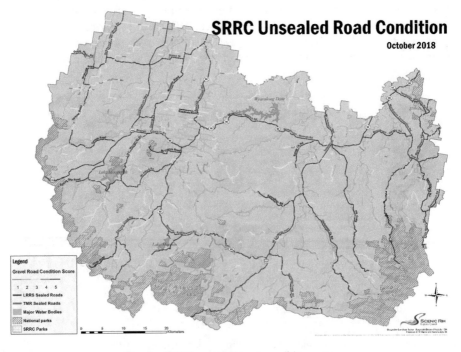

Fig. 4. SRRC unsealed road condition map.

Figure 4 shows the SRRC unsealed road condition map with colour coded condition rated roads. As this map is produced after the 70% completion of flood recovery works the whole network is in better condition. The network assessed in condition states requiring intervention usually reviewed with the high definition video imagery through the GIS or AMS by local works coordinators or managers who then prioritise and distribute maintenance resources having regard for their own judgement and experience and considering the operational circumstances at that time.

Visual Assessment has been a good tool on a network basis but it is based on surface condition and does not include gravel type. From Pavement and structural stability point of view there is need for better technique or assessment.

5 Gravel Loss Monitoring, Assessment and Expected Outcomes

SRRC has started resheeting and regravelling unsealed road network damaged during Ex Tropical Cyclone Debbie. SRRC has initiated gravel loss and maintenance strategy development project for unsealed road network. Under this project 56 gravel loss monitoring stations are established. There are 23 stations in the east region and 33 stations in the west region.

To reduce the resource demand and improve efficiency in data collection latest GIS App and QGIS software is used. Gravel loss monitoring will continue for 2 years and the Martin 2013 (ARRB) gravel loss model will be further refined for localised conditions. While completing the resheeting and regravelling the gravel material specification is amended and improved. The gravel material is sourced using this updated specification. Aim is to reduce gravel loss by use of this improved material. The material specification will be further developed during this study for wet and dry weather conditions. In addition to gravel loss measurement using gravel loss stations SSRC is measuring roughness on the section of roads where gravel monitoring stations are established. The aim of this roughness measurement is to establish any relationship between gravel loss and roughness on the road. The gravel loss at particular station provides cross section of the road over period of time. Roughness provides the roughness level along the road to driving condition.

6 Advantages and Limitations of Visual Inspections

Visual inspections provide snapshot of network. It is very good tool for Asset manager to demonstrate the overall condition of unsealed road network in a particular time frame. It provides condition rating of a particular road or a section of road. Visual inspections can be summarised to provide stakeholders and customers at what condition network is. For stakeholder or customer complaints it is possible to check for quick review as a desktop if not in a position to go and visit site. Ease to assess improvement or degradation of overall network based on summary report. Provided the inspection is carried out annually it is possible to work out if network is improving or degrading and what rate. It is good insurance policy in case of flood, safety incidents/accidents.

Even though the visual condition assessment is focused on surface characteristics and functionality of asset it is possible to use video imagery to check for safety incidents/accidents related to customer claims or investigations. Although author does not support this idea but at some instances it may be useful as a historical record.

Being visual inspection there is inherent limitation not able to provide overall condition (Subgrdae, drainage). Difficulty to develop routine maintenance program based on visual condition rating as this inspection identifies only surface defects gravel depths are not measured and similarly drainage problems. Due to rapidly changing characteristic of unsealed roads makes it difficult to justify usefulness of visual inspections.

Data collection and management of unsealed roads are focused on operational processes and lack of useful information at tactical and strategic level (Henning et al. 2019). 'Assessment Process for the Condition of Unsealed Roads, NZ Transport Agency research report' recommends further development performance reporting and system to assist with analysis for tactical levels. SRRC visual inspections provides higher level of maintenance strategy based on road lengths requiring immediate attention. Those areas can be further investigated in detail using other decision-making frameworks which can add value to the unsealed road management process.

7   Conclusion

It is difficult to predict unsealed road behaviour hence decision frameworks for unsealed roads are normally a combination of practical and theoretical approaches. Managing unsealed roads often involves operational issues, because unsealed roads change very quickly and when defects appear they must be addressed within a short response time. Many times, routine and cyclic maintenance is planned and scheduled according to routine inspections and experience from road operators.

However, longer-term maintenance activities, such as re-gravelling and surfacing of unsealed roads, need a more sophisticated process that includes predictive models. A major consideration during these analyses involves the economic appraisal of different maintenance options and timings of intervention.

Visual inspection can be a useful tool to provide snapshot of unseal road network at a time. Most road agencies still use this visual inspection as a maintenance strategy tool. It is recommended in this study to integrate visual inspection with localised gravel loss model and roughness data. GPS based data can be integrated by using visual inspections, roughness and the gravel model.

References

Australian Rural Roads Group: Going Nowhere: The rural local road crisis Its national significance and Proposed Reforms, Australian Rural Roads Group, NSW, Australia (2010)

Scenic Rim Regional Council (SRRC): 2018–19 Community Budget Report Scenic Rim Regional Council, Scenic Rim Regional Council, Beaudesert, Australia (2018)

Alzubaidi, H., Magnusson, R.: Deterioration and rating of gravel roads: state of the art. Road Mater. Pavement Des. **3**(3), 235–260 (2002)

McManus, K.J.: Pavement deterioration models for a local government authority. In Proceedings of 17th ARRB Conference, Part 4, Gold Coast, Queensland, 15–19 August 1994, vol. 17 (1994)

Lea, J.D., Paige-Green, P., Jones, D.: Neural networks for performance prediction on unsealed roads. Road Transp. Res. **8**(1), 57 (1999)

Henning, T.F., McCaw, A., Bennet, N.: Assessment process for the condition of unsealed roads. NZ Transport Agency research report 652 (2019)

Henning, T.F., Flockhart, G., Costello, S.B., Jones, V., Rodenburg, B.: Managing gravel-roads on the basis of fundamental material properties (No. 15-2562) (2015)

Henning, T., Giummarra, G.J., Roux, D.C.: The development of gravel deterioration models for adoption in a New Zealand gravel road management system (No. 332) (2008)

Martin, T., Choummanivong, L.: The benefits of Long-Term Pavement Performance (LTPP) research to funders. Transp. Res. Procedia **14**, 2477–2486 (2016)

Giummarra, G.: Unsealed Roads Manual: Guidelines to Good Practice, 2009th edn. Australian Road Research Board, Melbourne (2009)

Road Infrastructure Management Support (RIMS): Unsealed Roads Tactical Asset Management Guide, RIMS, New Zealand (2015)

Jones, D., Paige-Green, P., Sadzik, E.: Development of guidelines for unsealed road assessment. Transp. Res. Rec. J. Transp. Res. Board **1819**, 287–296 (2003)

Experimental and Numerical Investigations of Flexural Behaviour of Composite Bearers in Railway Switches and Crossings

Sakdirat Kaewunruen[✉], Pasakorn Sengsri,
and Andre Luis Oliveira de Melo

School of Engineering, The University of Birmingham, Birmingham, UK
s.kaewunruen@bham.ac.uk,
{PXS905,AL0888}@student.bham.ac.uk

Abstract. Composite bearers, which are the long crosstie beams, are safety-critical components in railway switches and crossings. Recent adoption of composites to replace aging timber bearers has raised the concern about their engineering performance and behaviour. Since the design and test standards for composite bearers are not existed, most performance evaluations are based on the flexural tests in accordance with the test standards for railway concrete sleepers. In this study, both numerical and experimental studies into the flexural behaviours of composite bearers have been conducted to improve the understanding into the resilience and robustness of the components under service load condition. The full-scale composite bearers are supplied by an industry partner. The full-scale tests have been conducted in structures laboratory at the University of Birmingham. 3D finite element modelling of the bearers has been developed using Strand7. The comparison between numerical and experimental results yields an excellent agreement with less than 3% discrepancy. The results exhibit that the composite bearers behave in the elastic region under service load condition. This implies that they can recover fully under the load, enhancing engineering resilience of the turnout systems.

1 Introduction

A novel composite material, 'fibre-reinforced foamed urethane (FFU)' has gained an important momentum for applications in railway industry. As railway bearers in switches and crossings, the FFU components acting as a beam are to redistribute the train forces (static and dynamic loads) onto track support (ballast). Also, they can secure the rail gauge to allow trains to travel safely [1–3]. Its structural performance must be instigated and assured at all time through inspection (safety-related assessment functions), monitoring (surveillance functions) and maintenance [4–6]. A further function of the structural elements in a ballasted railway track system is to aid lateral track protection to enhance the stability and stiffness of the track structure. Any structural deterioration or poor conditions of the elements could affect the reliability, safety, and quality of the railway line. This leads to impaired rail services, for example, if the bearers cracked dramatically, they would deform highly under the loads induced by wheel-rail interaction. This large differential settlement encourages the damage to

© Springer Nature Switzerland AG 2020
L. Mohammad and R. Abd El-Hakim (Eds.): GeoMEast 2019, SUCI, pp. 104–113, 2020.
https://doi.org/10.1007/978-3-030-34187-9_8

other railway elements that in turn shortens the maintenance period of the railway line. However, if the bearers are more flexible (low elasticity), the track can dramatically deform and providing the outcome in a large differential local track surface (top smoothness) [7–12]. These cause higher dynamic loads, poor travelling comfort and extra train energy consumption [13–15]. Additionally, if the lateral resistance of the line is inadequate to support horizontal loads, (i.e. due to loosened ballast or abraded bearers), rail buckling may occur [16].

Railway urban turnout is a unique track system employed to divert a train from a particular direction or a particular line onto other lines or other directions. It is a structural grillage system which comprises of steel rails, crossing (uncommon line elements), points (well-known as switches), rubber pads, steel plates, insulators, screw spikes, fasteners, beam bearers (either polymer, concrete, steel, or timber), ballast, and formation as shown in Fig. 1. Conventional turnout structural were typically supported by timber bearers. They allow the steelwork to be mounted directly on steel plates which are spiked or screwed into the bearers. Timber has an outstanding damping coefficient, whilst steel and concrete tend to have nearly no damping coefficient [17–20]. Concrete has proven to be a great counterpart to improve line and turnout stability – laterally, vertically [21, 22]. Moreover, steel bearers perform well in a short period, anyway, having higher turnout settlement and ballast breakage during the long period [23, 24].

Fig. 1. Typical turnout geometry [7].

Material scarcity and environmental concern force researchers considering new materials capable of satisfying the railway system specifications. Developing new materials capable of satisfying the functional requirements including enhancing their recyclability. There is a constant search for a material that is durable, reasonably easy to produce and maintain, has attractive costs, and meets the expected requests effectively [25]. A crucial concern in the railway industry is the replacement of deteriorated and damaged bearers in existing lines [25]. Especially in special positions such as railways crossings and switches, railway bridges, and transition zones, the requirement for alternative materials to replace old timber components is undoubtedly important [26,

27]. It is well known that common turnout generally imparts high impact actions on to structural members due to its blunt geometry and mechanical connections between closure rails and switch rails. This has boosted the importance of structural performance and failure modes of the elements employed in railway systems. Because the design method for composite bearers has not been standardized yet [1–4], most design concept for the composites is based on allowable stresses, which is slightly more conservative (more safety margin) compared with limit states design principle (Fig. 2).

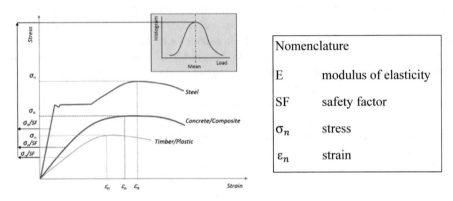

Fig. 2. Schematics of allowable stress design concept for each material [28].

As a result, it is imperative that material reduction factors be identified based on the data recorded through experiments into failure mode and structural performance. These in-depth understandings would later help the railway owner or authority to verify the cost effectiveness and safety possibility of the composite bearers. This allowable stress design concept determines the maximum strength of constituent materials, which then cannot be exceeded in the component. Safety and serviceability aspects such as brittle fracture, bursting, fatigue failure, and allowable deflections are taken into account in this design method by the determination of safety factor values [29]. The cost effectiveness can then be evaluated using reliability indexes whether the component is either optimally, overly, or under designed.

There are many efforts towards improving the characteristics of the materials already utilized in the railway track engineering (wood, concrete, and steel) as applied to the polymer by itself or composite polymers, using primarily fibres [30]. For over 35 years, fibre reinforced foamed urethane (FFU) composites have been utilized in the construction of railway track systems. Sekisui Chemical & Co [31] is the principal producer of this material. Numerous researches using Japanese testing standards are conducted for this material in order to define its limits of use or validated them in specific and particular cases [31]. On the other hand, based on the significant review for composites [1], it is clear that there is no previous work to evaluate the failure mode, structural damage and structural design performance of the FFU composites. The aim of this study is to focus the flexural behaviours of FFU composite bearers obtained from numerical and experimental data to enhance the insight into the resilience and robustness of the components under service load condition. The main highlight is to underpin the design ideas of plastic and FFU composite bearers. This is because the use of such bearers is relatively new in railway industry across the world. Knowledge of the engineering design principle is therefore significant for enabling proper repair, adoption, and retrofit of the line components in the future. In this paper, the experimental and numerical investigation into the flexural modes connected with plastic and composite bearers are presented. These understandings will aid railway engineers to determine suitable engineering methods and solutions for track construction and maintenance under future uncertainties.

2 Materials and Experiment of Composite Beams

An industry partner provided nine full-scale beams (160 mm depth × 250 mm width × 3200 mm length) using fibre reinforced foamed urethane composites (designed for railway track components). The experimental testing is based on the evaluation benchmark of EN 13230 (Test material specifications, support conditions, loading procedures, and some specific requirements for bending tests on railway track concrete sleepers). Some procedures are followed by these tests to prove the test information. On the other hand, EN 13230 has severe limitation in order to determine failure mode of flexible composites. For example, some test procedures are adjusted to investigate the structural damage and the failure mode of the full-scale FFU composite beams [32–34].

Positive and negative bending tests are needed at the rail seats support, based on EN 13230-2 bending testing. Since the FFU test specimens have the same positive and negative capacity (symmetry). In this paper, there is the only positive bending tests conducted to identify a resemble failure mode in a track system [25, 34, 35]. The standard requires articulated support and must be 100 mm wide, made of steel with Brinell: HBW > 240. In the experiments, quasi-static load should be applied to the middle span of the beam for normal positive bending. Figure 3 demonstrates the layout of the bending load process. Additionally, it indicates the locations of many different non-destructive test (NDT) sensors. Linear variable differential transformers (LVDT) are placed at the mid-span location of the specimen to collect the deflection. Also, three acoustic emission (AE) devices are employed in each test. Four strain gauges are set up at the front and rear locations of each specimen to record stress changes.

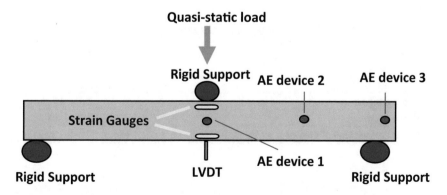

Fig. 3. Experimental setup of a full-scale FFU composite beam under flexural load.

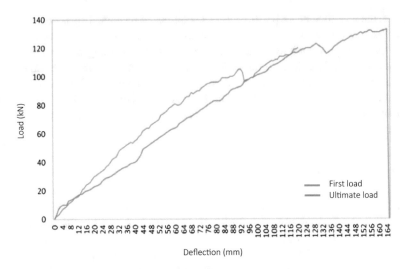

Fig. 4. Load-deflection behaviour of full-scale FFU composite beams under flexural load [28].

3 Finite-Element Model of Composite Beam

The finite-element model of an FFU beam under flexural load was developed to investigate its flexural behaviour. The sleeper model based on Timoshenko theory is the most agreeable approach for generating two-dimensional concrete sleepers [36–39]. Anyway, this model was modelled using 6,000 bricks with a trapezoidal cross-section in STRAND7 [40, 41], as shown in Fig. 5. Also, the finite element model consists of the brick components, which take into account flexural and shear deformations, in order to model the FFU beam model. The material and geometric properties of these bricks are presented in Table 1. These properties were chosen because they were identical to a particular type of bearers manufactured in the UK. A flexural mode shape was conducted to evaluate the quality of the finite element (FE) model. It was found that 6,000 bricks, representing a composite beam, can provide acceptable estimation of bearer's flexural behaviours under service load condition compared with the existing experimental measures.

Table 1. Engineering properties used in the modelling.

Parameter lists	Value	Unit
Elastic modulus	7000	MPa
Poisson's ratio	0.25	–
Beam density	600	kg/m^3
Beam length	3.2	m
Beam cross-sectional area of a rectangle (0.16, depth * 0.25, width)	0.042	m^2

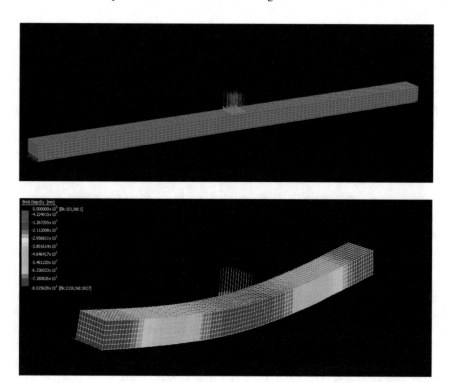

Fig. 5. Finite element analysis for modelling and behaviour of an FFU composite beam under flexural condition.

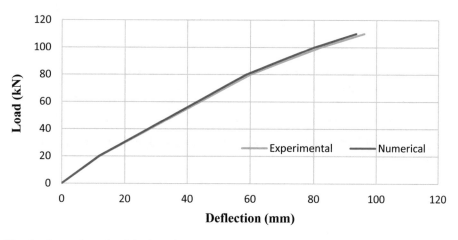

Fig. 6. Comparison the defection of Flexural behaviour between the numerical and experimental data under flexural load

In order to verify the model, the flexural bending mode shape of a beam under static load condition obtained from the FE model were compared with experimental data. Figure 6 shows a comparison between the finite-element analysis and experimental results of deflection. The outcomes are found to be in a very good agreement with around less than 3%, as given in Fig. 6. However, this study shows the comparison between the FEA and experimental data in the only static linear analysis under flexural behaviour in first load condition.

4   Result and Discussion

As shown in Fig. 4, crack propagation has been collected and marked under the load increment. It is obvious that the first crack (fracture of internal fibres) occurs at 34 kN and composites failed at 132 kN. Figure 7 demonstrates the crack propagation appears. The fibre cracks can be observed longitudinally along the fibre orientation. Also, minor fractures can be noticed before the sudden failure of the composite beams. Figure 8 illustrates the failure mode of the composite beams in laboratory. Obviously, first cracks or localized fibre failure is relative low, compared with the ultimate failure load. The rapid rupture can be seen through the delamination of fibres along with the beams. In fact, the brittle failure mode of the composites controls as the consideration of the load deflection given Fig. 4. In addition, when larger fibre breaks are seen, the composites behave nonlinearly towards the failure point. While the linearization of structural behaviour is true only when the small fibre breaks occur.

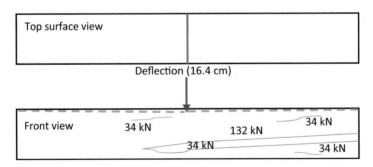

Fig. 7. Crack behaviour of full-scale FFU composite beams under flexural load [28].

Fig. 8. Failure mode of full-scale FFU composite beams under flexural load [28].

5 Conclusion

Flexural behaviours of Fibre-reinforced foamed urethane (FFU) bearers in railway system are crucial for improving the understanding into the resilience and robustness of the components under service load condition. The flexural behaviours of an FFU composite model were studied using the finite element approach. Whilst, FFU beam specimens used in the experimental test were conducted employing a three-point bending test. The experimental investigations into the failure modes of FFU composites. The new insight into the load defection, crack propagation, and failure mode will aid the rail industry to produce a better decision for proper adoption of composites in railway infrastructure. The three-dimensional modelling has been verified and found in very good agreements with the experimental data with less than 3% difference, as given in Fig. 6. According to the use of standard test approaches, it is confirmed that existing standard cannot be acceptable for the composite materials. Obviously, the results of flexural behaviours between numerical and experimental data cannot be validated without better understandings of in-track behaviours, failure modes, and science-based design approach for the materials. Additionally, this paper identifies that the composites are likely to have brittle failure modes. This means that a more safety factor should be applied for the element approach. The rupture cannot literally be observed from the progression of cracks appeared on the surface of the element. The development of condition monitoring implement is essentiality before wide-spread use of composites in railway industry. This model has been very useful and has led to further research on structural behaviours of the railway FFU bearers in the track structure system or as well known 'the in-situ railway FFU bearer'. Also, a numerical analysis under failure condition should be conducted to absolutely verify the experimental data.

Acknowledgments. The authors would like to acknowledge to the European Commission for the financial sponsorship of the H2020-MSCA-RISE Project No. 691135 "RISEN: Rail Infrastructure Systems Engineering Network," which enables a global research network that tackles the grand challenge in railway infrastructure resilience and advanced sensing in extreme environments (www.risen2rail.eu) [42]. In addition, this project is partially supported by the European Commission's Shift2Rail, H2020-S2R Project No. 730849 "S-Code: Switch and Crossing Optimal Design and Evaluation".

References

1. Silva, E.A., Pokropski, D., You, R., Kaewunruen, S.: Aust. J. Struct. Eng. **18**, 160–177 (2017)
2. Kaewunruen, S.: J. Compos. Constr. **23**, 07018001 (2019). https://doi.org/10.1061/(ASCE) CC.1943-5614.0000833
3. Kaewunruen, S., Remennikov, A.M., Murray, M.H.: Front. Mater. **1**, 8 (2014). https://doi. org/10.3389/fmats.2014.00008
4. Kaewunruen, S.: Struct. Monit. Maint. **1**(1), 131–157 (2014)
5. Kaewunruen, S.: Case Stud. Nondestruct. Test. Eval. **1**(1), 19–24 (2014)
6. Kaewunruen, S., Bin Osman, M.H., Rungskunroch, P.: Front. Built Environ. **4**, 84 (2018). https://doi.org/10.3389/fbuil.2018.00084
7. Kaewunruen, S., You, R., Ishida, M.: Composites for timber-replacement bearers in railway switches and crossings. Infrastructures **2**, 13 (2017). https://doi.org/10.3390/ infrastructures2040013
8. Kaewunruen, S.: Proceedings of 2013 International Workshop on Railway Noise, 9th–13th September 2013, Uddevalla, Sweden (2013)
9. Kaewunruen, S., Remennikov, A.M.: Eng. Fail. Anal. **16**(3), 705–712 (2009)
10. Gamage, E.K., Kaewunruen, S., Remennikov, A.M., Ishida, T.: Infrastructures **2**, 3 (2017). https://doi.org/10.3390/infrastructures2010003
11. Gamage, E.K., Kaewunruen, S., Remennikov, A.M., Ishida, T.: Infrastructures **2**, 5 (2017). https://doi.org/10.3390/infrastructures2010005
12. Tavares de Freitas, R., Kaewunruen, S.: Environments **3**, 34 (2016)
13. Kaewunruen, S., Martin, V.: Sustainability **10**, 3753 (2018)
14. Remennikov, A.M., Kaewunruen, S.: Struct. Control. Health Monit. **15**(2), 207–234 (2008)
15. Kaewunruen, S., Remennikov, A.M.: Aust. J. Civ. Eng. **14**, 63–71 (2016)
16. Bin Osman, M.H., Kaewunruen, S.: Proc. Inst. Mech. Eng. Part F J. Rail Rapid Transit (2019)
17. Kaewunruen, S.: Monitoring in-service performance of fibre-reinforced foamed urethane material as timber-replacement sleepers/bearers in railway urban turnout systems. Struct. Monit. Maint. **1**, 131–157 (2014)
18. Kaewunruen, S.: In situ performance of a complex urban turnout grillage system using fibre-reinforced foamed urethane (FFU) bearers. In: Proceedings of the 10th World Congress on Rail Research, Sydney, Australia, 25–28 November 2013
19. Kaewunruen, S.: Monitoring structural deterioration of railway turnout systems via dynamic wheel/rail interaction. Case Stud. Nondestruct. Test. Eval. **1**, 19–24 (2014). (CrossRef)
20. Indraratna, B., Salim, W., Rujikiatkamjorn, C.: Advanced Rail Geotechnology—Ballasted Track. CRC Press/Balkema, Leiden (2011)
21. Remennikov, A.M., Kaewunruen, S.: A review of loading conditions for railway track structures due to train and track vertical interaction. Struct. Control. Health Monit. **15**, 207–234 (2007). (CrossRef)
22. Standards Australia: Australian Standards: AS3818.2 Timber; Standards Australia, Sydney, Australia (2001)
23. RailCorp: Timber Sleepers & Bearers; Engineering Specification SPC 231; RailCorp, Sydney, Australia (2012)
24. Kaewunruen, S., Remennikov, A.M.: Dynamic flexural influence on a railway concrete sleeper in track system due to a single whel impact. Eng. Fail. Anal. **16**, 705–712 (2009). (CrossRef)

25. Van Erp, G., McKay, M.: Recent australian developments in fibre composite railway sleepers. Electron. J. Struct. Eng. **13**, 62–66 (2013)
26. Dindar, S., Kaewunruen, S., An, M., Sussman, J.M.: Saf. Sci. **110**, 20–30 (2018)
27. Freimanis, A., Kaewunruen, S.: Appl. Sci. **8**(11), 2299 (2018)
28. Kaewunruen, S., Goto, K., Xie, L.: Experimental investigations into failure modes of fibre reinforced foamed urethane composite beams. In: Selection and Peer-Review Under Responsibility of 4th Advanced Materials Conference 2018, 4th AMC 2018, 27th & 28th November 2018, Hilton Kuching Hotel, Kuching, Sarawak, Malaysia (2018)
29. Kaewunruen, S., Remennikov, A.M.: Electron. J. Struct. Eng. **13**, 41–61 (2013)
30. Manalo, A.: Behaviour of fibre composite sandwich structures: a case study on railway sleeper application. Ph.D. Thesis, Centre of Excellence in Engineered Fibre Composites Faculty of Engineering and Surveying University of Southern Queensland Toowoomba, Queensland, Australia (2011)
31. Sekisui Co.: Engineering Properties of FFU Materials. Sekisui Co., Tokyo, Japan (2012)
32. Li, D., Selig, E.: Evaluation of railway subgrade problems. Transp. Res. Rec. **1489**, 17 (1995)
33. Tata Steel: Steel Sleepers, 1st edn. Tata Steel Europe Ltd., Brockhurst Cres, Walsall, UK (2014). http://www.tatasteeleurope.com. Accessed 20 Apr 2019
34. Health and Safety Executive: Rail Track and Associated Equipment for Use Underground in Mines (2007). http://www.hse.gov.uk/pubns/mines06.pdf. Accessed 23 Apr 2019
35. European Federation of Railway Trackworks Contractors: Newsletters EFRTC (2007). http://www.efrtc.org/htdocs/newsite/newsletters.htm. Accessed 1 May 2019
36. Cai, Z.: Modelling of rail track dynamics and wheel/rail interaction. Ph.D. thesis, Department of Civil Engineering, Queen's University, Ontario, Canada (1992)
37. Grassie, S.L.: Dynamic modelling of concrete railway sleepers. J. Sound Vib. **187**, 799–813 (1995)
38. Kaewunruen, S., Remennikov, A.M.: Sensitivity analysis of free vibration characteristics of an in situ railway concrete sleeper to variations of rail pad parameters. J. Sound Vib. **298**, 453–461 (2006)
39. Neilsen, J.C.O.: Eigenfrequencies and eigenmodes of beam structures on an elastic foundation. J. Sound Vib. **145**, 479–487 (1991)
40. G+D Computing: Using STRAND7 introduction to the Strand7 finite element analysis system, G+D Computing Pty Ltd. (2002)
41. G+D Computing: STRAND7 theoretical manual, G+D Computing Pty Ltd. (2005)
42. Kaewunruen, S., Sussman, J.M., Matsumoto, A.: Front. Built Environ. **2**, 4 (2016). https://doi.org/10.3389/fbuil.2016.00004

Information Communication Technologies for Travel in Southern African Cities

N. Bashingi[1](✉), M. Mostafa Hassan[2], and D. K. Das[3]

[1] Sustainable Urban Roads and Transportation research group,
Department of Civil Engineering, Central University of Technology, Free State,
Bloemfontein, South Africa
nbashingi@gmail.com
[2] Civil Engineering, School of Engineering,
University of KwaZulu-Natal, Durban, South Africa
MostafaM@ukzn.ac.za
[3] Department of Civil engineering, Central University of Technology, Free State,
Bloemfontein, South Africa
ddas@cut.ac.za

Abstract. This paper investigates the current state of ICT use for travel by public and private transportation users in two different Southern African cities of Bloemfontein, South Africa and Gaborone, Botswana. ICT use is seldom in transportation systems in the developing world, the transportation systems are still conventional with minimal communication technologies supporting the system. This study investigates ICT use for travel and travel purposes for which ICT is currently applicable.

Before investigating ICT use, the study first establishes the need for travel and modal split between activities, then explores ICT knowledge and access in both cities to determine feasibility of promoting ICT for travel. The current state of ICT usage for transportation and travel purposes is investigated to determine ICT influence on travel. The study further investigates the perceptions of transportation users towards ICT in transport by investigating ICT components (devices, applications and software) which would be useful for travel at different stages of a trip, and the purposes for which the users would use them.

Keywords: ICT · Public transportation · Travel behaviour · User perception

1 Introduction

Transportation in an urban areas is one of the most important systems for the area's efficiency. Mobility of people, goods and services through the transportation systems enables functionality between all of the city's systems. However, most of the world's cities have not been able to efficiently operate human travel systems that satisfy the urban area. Transportation solutions evolved from creation and development of new models of passenger cars, buses and trains to searching for solutions in information technologies. Emergence of Information Communication Technologies (ICT) prompted their exploitation by various industries as catalysts to communication problems.

L. Mohammad and R. Abd El-Hakim (Eds.): GeoMEast 2019, SUCI, pp. 114–127, 2020.
https://doi.org/10.1007/978-3-030-34187-9_9

ICT was catapulted into transportation through logistics and traffic monitoring; the use of Automatic Vehicle Location and Intelligent Transportation Systems.

Advancement of technology and high speed internet brought forth the proliferation of mobile communication devices, smartphones and mobile applications which have contributed to change in travel behaviour and patterns. Recent impact of ICT on transport, which greatly impacted travel and travel behaviour is the surge in ride sharing platforms. Technologies developed from GIS software to Global Positioning Systems (GPS) to enabling simplified navigation and mapping application accessible through mobile devices. Ridesharing platforms leveraged the growth of ICT, created mobile applications providing travellers with more convenience and flexibility though increased travel mode options as well as methods of payment for travel.

Larger cities in Southern Africa such as Johannesburg and Cape Town have advanced public transportation systems compared to Bloemfontein and Gaborone. High speed trains, public transportation travel applications such as GoMetro and WhereIsMyTransport, as well as ride sharing platforms like Uber and Lyft are available in these cities.

Impacts of ICT on travel behaviour have also been researched over the years, including the use of ICT as a substitution to travel. The penetration of ICT in transport however did not occur universally at the same time. Developing countries are still behind on the technological trends. African countries, faced with high poverty levels did not prioritise adoption of ICT. Hinderance of ICT growth is also influenced by the high costs of ICT devices, high cost of internet, which leads to lack of access to ICT. African countries also face severe lack of ICT infrastructure, which also contributes to the high cost of internet access.

2 Study area

The study was conducted in two Southern African cities, Bloemfontein in South Africa and Gaborone, Botswana. Bloemfontein, which includes the Mangaung area has a population of 463 699 (World Population Review 2017), while the city of Gaborone has a population of 232 000 which is predicted to grow to 250 000 by 2021 (Statistics Botswana 2016).

Although Gaborone is the capital city of Botswana and Bloemfontein is the capital city of the Free State province in South Africa, they are relatively smaller, in population size, geographic characteristics and urban structure compared to some of the major cities in the Southern African region. Transportation systems in both cities are dominated by private vehicle use and the use of traditional public transportation systems. Public transportation system is composed of a bus service by a single operator, Interstate Bus Lines, mini-bus taxis, 5 seater taxis as well as cabs operated by various private companies and individuals. Gaborone's public transportation system is comprised of mini-bus taxis (known as combis), 5 seater taxis (commonly known as a taxi) and on-demand cabs. Application based rise sharing platforms are not available in both these areas.

3 Literature review

ICT's influences on transportation include impact on physical transportation structures as well as organisation of transportation services. ICT also influences travel demand and preferences (Snellen and De Hollander 2017). Information Communication Technologies have greatly impacted the way people travel. Travel behaviour and patterns, dissected into mode choice, time of travel, route choice, travel activities and reason for travel are influenced by access or lack of ICT thereof. ICT and transportation can influence activity travel pattern by modification, substitution and activity generation (Lila and Anjaneyulu 2016). Dissemination of ICT devices and internet access supports activity fragmentation and multitasking during travel (de Abreu e Silva et al. 2017). ICTs are changing demand for mobility and tools to meet the demand for mobility, therefore, ICT has the ability to change transport systems and improve travel experiences (Snellen and De Hollander 2017).

Previous studies have resulted in four main impacts if ICT on travel, ICT substituting travel, ICT complementing travel, ICT neutralising travel and ICT modifying travel (Mokhtarian 1990). ICT in transportation has the following change effects; more information is now available, it increases the range of transport options available to users, reduced the need to travel, changes perceptions on travel time, geography of destinations change, the mobility system complexity increases and finally the arrival of self-driving vehicles (Snellen and De Hollander 2017).

The purpose of travel is primarily to access activities but with growth in technology, some activities have been virtually replaced. Lila and Anjaneyulu (2016) explored the impacts of teleworking, e-commerce, tele-medicine on transportation. Teleworking or working from home has increased majorly since the advancement of ICT and has been seen to reduce congestion and energy consumption as a result of reduced home to work trips (Melo and Silva 2017). Even though ICT has the potential to reduce travel, it also may largely increase travel as it improves flexibility to daily travel and travel decisions ((Ben-Elia et al. 2014).

The benefits that ICT could bring to transport are linked to the ability and willingness of people to change their travel behaviour (Snellen and De Hollander 2017). Studies also show that travellers weigh their options before beginning their trips, as a way of managing uncertainties, (Nyblom 2014), although, the most important part of travel planning is thinking ahead. The use of information from analogue, informal information services and ICT based information is common for trip planning. Besides the use of ICT to access information before journeys, ICT is useful during and after trips.

Emergence and improvement of portable and mobile technologies have also impact travel behaviour and patterns, specifically decision making. Portable ICT components such as cell phones, laptops, tablets as well as wireless internet connections such as mobile data and portable Wi-fi have enabled virtual performance of activities from anywhere (Ben-Elia et al. 2014). Multitasking during trips by form of performing virtual activities has risen due to access to ICTs. The ability to perform multiple tasks have increased with the increase in availability of ICTs (Varghese and Jana 2018).

Availability and access of ICT is not homogenous world-wide due to different countries socio-economic aspects. Socio-economic status in developing countries impacts ICT knowledge and access, financial status as well as living standards influence access and types of devices affordable to users. Socio-economic factors also influence mode choice as well as reasons for travelling (Li and Lo 2018). Factors such as age and duration of ICT use can also influence virtual mobility, adaptation and use of ICT and travel (Konrad and Wittowsky 2018).

Most studies focus on ICT as a way of decreasing the need to travel and substituting travel, focusing on spatial, time and activity relationships, however, the use of ICT for the purposes of travel itself are not fully investigated. The relationship between ICT and travel purposes have not been thoroughly researched (de Abreu e Silva et al. 2017). It is evident that ICT would be impactful to travel, but there is no clear indication of purposes for which ICT is used.

4 Methodology

Research data was acquired using face to face questionnaire surveys carried out on 415 individuals in Bloemfontein, South Africa and 388 individuals in Gaborone, Botswana from August 2017 and November 2017. The participants of the study were both private and public transportation users.

The methodology aimed at exploring the state of ICT for travel and ICT use for public transportation as well as the possibilities of ICT use for travel purposes.

1. What is the state of ICT use for travel?
2. Is there scope for ICT incorporation into travel?

Modes of transportation used to access different activities were explored to establish travel characteristics of individuals. Variables used to measure the existing state of ICT use for travel purposes are (i) ICT use for public transportation and (ii) ICT components used for travel as well as (iii) the purposes of which ICT components are used.

To forecast the possibilities of future ICT use and its probable impacts on travel, the following variables were used, (i) ICT knowledge (ii) Internet access (iii) Activities during trip (iv) Importance of ICT before, during and after trips and lastly (v) the purposes of which ICT would be used before, during and after trips.

Diverse groups of respondent in both cities were used, their demographic characteristics are presented on Table 1. All the participants were asked the following questions:

1. Do you have any knowledge of ICT?
2. Do you have internet access?
3. Do you use ICT for any reasons related to public transportation travel?
4. Which of the following activities do you perform during your trip?
5. Which of the following ICT components do you use for travel?
6. At what stage of travel would ICT be useful for your travel?
7. For what purpose would ICT be useful before/during or after a trip?

Table 1. Demographic characteristics of study sample

Characteristics		Gaborone (%)	Bloemfontein (%)
Age	18–25	43,0	34,7
	25–35	37,4	40,2
	36–45	15,5	16,9
	46–55	4,1	7,2
	55+	0	1,0
Gender	Male	46,6	54,5
	Female	53,4	45,5
Occupation	Student	33,0	29,2
	Part-time employee	7,5	13,5
	Self employed	11,9	7,0
	Full-time employee	41,0	38,3
	Unemployed	5,9	6,7
	Student and Part-time employee	,8	5,3

5 Results

5.1 Travel Characteristics

First, we explore the need to travel as well as the mode of travel. Travel is generated by the need to perform activities. To fulfil the need to travel, travellers use various modes of transportation. Tables 2 and 3 show the modal split among activities.

Table 2. Modal split for activities in Bloemfontein

	Bus	Taxi	Cab	Private vehicle	Motorcycle	Bicycle	Walking
Shopping	7.6	61.1	19.5	17.3	0.7	1	34.5
Work	10.2	37.8	2.9	9.8	0.5	0.2	14.1
School	4.6	29.8	3.4	8	0.7	0.5	23.7
Personal	4.9	46	10.7	18.2	0.5	0.7	38.9
Leisure	2	27.7	12.4	14.6	0.7	1.7	28.2
Sporting activities	1.7	24.6	2.2	12.9	0.5	1.7	28.3
Sightseeing	1.7	19.8	3.7	13.2	0.5	1.2	26.8
Family obligations	8.3	39.3	3.4	23.4	0.7	1.2	17.5

On Table 2, which shows modal split in Bloemfontein, an overwhelming majority of shopping trips are by taxi, (61.1%), followed by 34.5% walking trips; 19.5% cab trips and 17.3% shopping trips by private vehicles. Only 7.6% of the trips are by bus. Work trips are mostly be taxi, (37.8%), while 14.1% work trips are by walking; 10.2% bus trips and 9.8% work by private vehicles. Further, 2.9% work trips are by cabs. Only 0.5 and 0.2% of trips to work are by motorcycles and bicycles consecutively. School or educational trips are 29.8% taxi trips and 23.7% walk to school. 4.6% of students take

Table 3. Modal split for activities - Gaborone

	Combi	Taxi	Cab	Private vehicle	Motorcycle	Bicycle	Walking
Shopping	44,2	11,1	11,6	41,3	0	0	16,2
Work	31,8	4,1	1,6	28,4	0	0	4,4
School	35,4	6,2	2,6	15,8	0	0	6,5
Personal	48	9,3	9,3	36,4	0,3	0	11,9
Leisure	35,7	6,2	15	39,8	0,5	0,5	11,9
Sporting activities	20,4	2,8	1,6	23,8	0,5	0,8	13,4
Sightseeing	15,2	1,6	2,1	36,2	0,3	0,3	15,2
Family obligations	30,7	6,2	6,7	51,9	0	0	3,9

the bus to school while 3.4% take a cab. Only 0.8% of respondents take private vehicles to school. Motorcycles and bicycles were the least common modes of transportation for travelling for all activities.

Combis and private vehicles, as shown on Fig. 3, are the most used modes of transportation in Gaborone. 44. 2% of respondents used combi for shopping trips while 41.3% used private vehicles A further 16.2% walked for their shopping trips; 11.6% used cabs while 11.1% used taxis. None of the respondents used motorcycles and bicycles for shopping trips. Travel to work in Gaborone is dominated by combi and private vehicles. 31.8% of respondents used combis for work trips while 28.4% used private vehicles. Only 4.4% and 4.1% walked and used taxis consecutively. School trips were also mostly dominated by combis and private vehicles; 35.4% combi trips; 15.8% private vehicles; 6.2% taxi trips and 2.6% cab trips. Personal trips, these are trips to perform miscellaneous activities not listed on the questionnaire. 48% of these trips are undertaken using combis, 36.4% by private vehicles; 11.9% walking trips and lastly 9.3% by taxis and cabs respectively. Travel for leisure, sporting, sightseeing and travel for family obligation activities indicated that majority of respondents used private vehicles over combis for these activities; 39.8% private vehicles over 35.7% combi leisure trips; 23.8% private vehicle sporting trips over 20.4% combi trips; 36.2% private vehicle sightseeing trips compared to 15.2% combi trips and lastly, 51.9% private vehicle use for family obligations compared to 30.7% combi use. Cabs are mostly used for leisure trips compared to other activities; 15% cab use for leisure trips; 6.7% cab trips for family obligations.

Travel for leisure activities is lagging behind travel for socio-economic activities such as education, work and shopping. The purposes of which trips are undertaken to fulfil helps determine applicability of ICT. Whether these activities could be substituted by ICT, whether their integration with ICT would create more reason to travel or could be modified by availability of ICT.

Table 4 shows activities performed by respondents simultaneously with their trips. Social networking platforms are the popular activities, during travel. Most of the activities are not related to travel itself but communication and entertainment. Performance of these activities is not in any way related to the trip but are for entertainment and personal communication purposes. Activities performed simultaneous to trip making are however limited to in-vehicle activities that are supported by access or ownership of technological devices or conventional media. Reading, conversations with other passengers, social networking using smartphones, entertainment such as music and gaming.

Table 4. Activities performed during trip

Bloemfontein				Gaborone			
Activity	Responses		Percent of Cases	Activity	Responses		Percent of Cases
	N	Percent			N	Percent	
Reading	21	3.9%	6.6%	Reading	17	2.3%	5.5%
Email	22	4.1%	6.9%	Email	38	5.1%	12.3%
Facebook/Twitter	97	18.1%	30.3%	Facebook/Twitter	229	30.5%	74.4%
WhatsApp	207	38.7%	64.7%	WhatsApp	264	35.2%	85.7%
Phone Call	82	15.3%	25.6%	Phone call	112	14.9%	36.4%
Games	20	3.7%	6.3%	Games	30	4.0%	9.7%
Music	13	2.4%	4.1%	Music	31	4.1%	10.1%
No activities	73	13.6%	22.8%	No activities	29	3.9%	9.4%
Total	535	100.0%	167.2%	Total	750	100.0%	243.5%

From the results, it is evident that activities accessible through mobile devices are common during travel. This is partially due to portability of mobile devices and wireless internet.

5.2   ICT Characteristics

Figure 1, shows ICT knowledge, access and use of ICT for public transportation related purposes. Both cities have high levels of internet access and ICT knowledge. 91.1% in Gaborone had access to the internet, and 89.1% in Bloemfontein and 82.3% of ICT knowledge in Gaborone and 81.7% in Bloemfontein. The study further investigated ICT use for public transportation reasons, 50% of respondents in Bloemfontein used ICT for public transportation reasons compared to 31.8% of respondents who used ICT for public transportation in Gaborone. Majority of respondents in both study areas have ICT knowledge and Internet access but do not use ICT for public transportation travel reasons. For ICT to be impactful towards travel, there need to be acknowledgement of its accessibility, availability as well as usefulness. Knowledge of ICT and access to internet as significant indicators of ICT penetration in to Southern African cities was assessed. The internet is the cornerstone to the functionality of various communication technologies, therefore, its access could determine the potential penetration of technology.

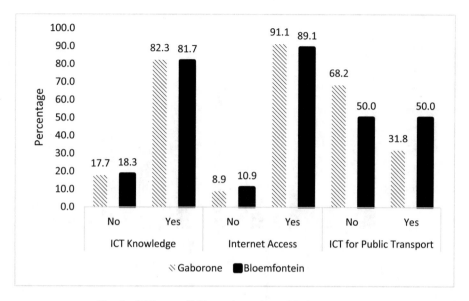

Fig. 1. ICT accessibility and use for public transportation

5.3   ICT Components for Travel

Figure 2, shows ICT components used for travel in Gaborone and Bloemfontein. There is high usage of devices, applications and ICT systems in in Bloemfontein compared to Gaborone. However, in both areas, smartphones, phone calls, SMS, social networking and web browsing are the mostly used. Purposes for which the devices were used for, the results indicated that the use of smartphones, for travel is mainly for social networking and entertainment purposes. The devices, applications and systems do not greatly influence trip making decisions and behaviour. Exceptions lie on the conventional use of phone calls and SMS communication between cab users and cab service providers. Cab trips, begin with the prospective passenger calling a cab company or driver to request their service. Communication thereafter relies on phone calls for further details, cancellations, notification of delays etc. Payment systems, through online platforms or smart card electronic systems are not available in both cities. Therefore, public transportation fare payments are made in the form of cash and fare loaded bus tags in Bloemfontein. Internet use for traffic updates and navigation are mostly used by private vehicle users.

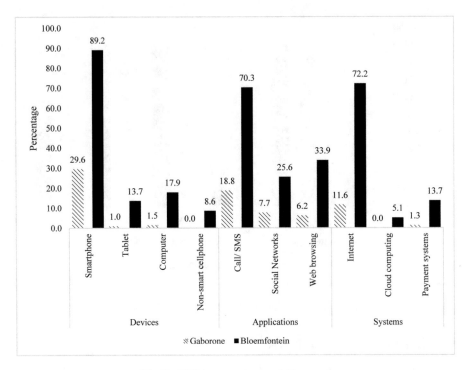

Fig. 2. ICT components used for travel

5.4   The Future of ICT and Travel

Figure 3, shows ICT potential usefulness at various stages of a trip. The results indicate that ICT would be useful for different purposes at three stages of a trip; Before, during and after trip completion. Majority of individuals in both study areas would be willing to use ICT for their trips. 91.6% of respondents in Bloemfontein and 88.3% in Gaborone would use ICT before their trips, 79.6% and 84.7% during the trip and finally 78.3% and 72.5% of respondents would use ICT after the trip. There is lower interest for ICT use after the trip compared to before and during trips.

Further, the purposes of which ICT would be used for were explored. Using a 5 point importance Likert scale of importance, respondents were requested to indicate the level of importance which they though various purposes would be at varying stages of the trip.

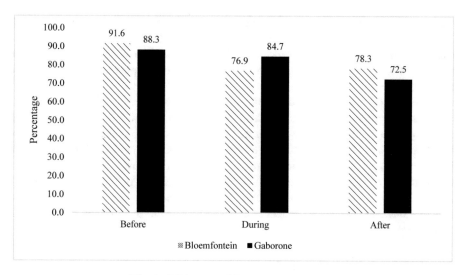

Fig. 3. ICT use at different stages of a trip

5.5   ICT Importance Before Trips

Uses of ICT before a trip included trip planning, inquiries, route mapping, sourcing information before a trip, comparing available modes of transport and for traffic information were provided. 20.4% and 55.4% of respondents indicated in Bloem-fontein that ICT would be important and very important to trip planning, 32.4% and 40.4% indicated that ICT would be important and very important for inquiries, 31.6% and 37.9% for route mapping, 25.5% and 40% for sourcing information before the trip, 25.4% and 35.1% for comparing modes of public transportation and 28.1% and 38.5% for traffic information before trips.

In Gaborone, 34.5% and 37.1% ICT would be important and very important to trip planning, 30.4% and 39.2% for inquiries, 36.1% and 34.5% for route mapping, 30.4% and 32.2% for destination information, 30.6% and 31.7% for travel mode comparisons and 29.1% and 29.9% for traffic information. Less than 20% of respondents showed ICT would not be important for all purposes.

Information is very essential in decision making, therefore, it is very critical for travellers to make informed decisions about their trips. People develop anxieties due to lack of information, lack of insight or knowledge about availability of the next available public transportation vehicle could be lessened by information provision. Purposes for which ICT could be used before trips are mostly related to reliability and convenience. Trip planning provides organisation before one embarks on a journey. Choosing the mode of transportation based on costs and time. Where available, users are able to compare all the available modes and schedules as well as fares, allowing them make travel decisions best suiting their travel needs.

Route mapping would be of more use to private transportation users whose flexi-bility permits them to travel routes they chose, make stops and at their own times. The use of ICT to map routes would not necessarily have the same impact on public transportation users. Public transportation routes are rigid and pre-set by respective Taxi Associations and the Department of Transportation. Public transportation users therefore would not have the freedom to plan a route and travel times. However, the ability to plan routes, for multiple trip journeys, users could plan routes based on the available modes for each trip.

Travel mode comparisons, traffic information although both forms of information, each has a significant influence on travel and decision making. Public transportation users compare all the available public transportation modes based on their schedules and fares before the trip. Although traffic congestion does not exclusively impact private vehicle users, access to traffic information would be most useful to them, Public transportation drivers and users do not have decision making powers to make choices relating to traffic and congestion. Public transportation routes are fixed, therefore, in case of traffic congestion, one cannot legally make decisions to avoid the traffic such as changing routes.

5.6   ICT Importance During Trips

ICT based activities that would be be performed during trips include trip planning, travel monitoring, entertainment, real-time traffic monitoring, work and duration monitoring.

Use of ICT for trip planning in Bloemfontein is not important to 28.2%, important and very important to 21.6% and 25.7% consecutively. Using ICT for travel monitoring was important and not important to 26.5% respectively and not important to 24.5%. Entertainment was very important to 28.4%, important to 21.8% and not important to 20.9%. Real-time traffic updates during the trip were very important to 29.4%, important to 24% and not important to 19.9%. ICT would be important for work to 23.3%, very important to 28.9% and not important to 23.1%. Monitoring the duration of trips would be important to 22.4%, very important to 27% and not important to 27%.

Results from Gaborone show that ICT for trip planning would be important to 28.8% and very important to 24.9% and not important to 26.8%. Travel monitoring is important and not important to 29.9% and 27.8% consecutively and not important to 20.5%. Entertainment during the trip was important to 28.3%, very important to 22.3% and not important to 18.2%. Real time traffic updates during the trip were important to 28.3%, very important to 21.3% and not important to 21.3%. ICT for work purposes would not be important to 27.3%, important to 22.6% and very important to 19%. Trip monitoring using ICT is not important to 22.6%, important to 31.7% and very important to 22.1%.

Planning during the trip does not carry much importance as before the trip begins. However, trip planning for the next trip for multiple-trip journeys might occur during the preceding trip. During the course of the trip, ICT could be useful for performing technology based activities for travel, work or entertainment. Travel related activities include planning the remainder of the trip, monitoring travel and traffic. When the right technology is available, passengers are able to perform work activities during the trip. Real time traffic updates are important, especially to private vehicle users who have the opportunity to change routes and travelling times as they are in control of their vehicles. Flexibility of changing routes based on real-time traffic updates has less impact for public transportation users without control of the vehicle. Working during the trip would be useful for people travelling longer journeys as passengers. Laptops, tablets, smartphones and availability of internet aboard vehicles could enable travellers to perform work or school related activities during their trips. This however, given the state of safety in public transportation systems could raise security concerns. ICT accessible to drivers for work reasons are phone call functions connected to vehicle audio systems through Bluetooth technology. Entertainment activities such as music, watching videos, video games also have the potential to attract more passengers to public transportation, especially younger travellers. This could change adult drivers travel patterns as often they have multiple trips to cater for younger family members activities such as school before undertaking their own trips.

5.7   ICT Importance After Trips

Bloemfontein results show that ICT for payment after trips was very important to 40.5% of respondents and not important to 22.4%, while the use of ICT for trip evaluation at the end of the trip was not important to 27.1%, important to 26.6% and very important to 23.2%. Using ICT for service evaluation after trips was not important to 28.4%, important to 20.8% and very important to 25.7%. Use of ICT to log queries was very important to 26.1%, important to 21.7% Rating trips by public transportation users would not be important to 32%, very important to 28.8% and important to 17.1%.

Results from Gaborone indicated that use of ICT for payments in after trips is very important to 35.8%, important to 26.2% and not important to 24.4%. ICT for trip evaluation is important to 32.2% and not important to 30.6%; Very important to 14%, Fairly important to 13.2% and slightly important to 3.6%.

Use of ICT to evaluate services after trips is not important to 30.4%, important to 29.9%, fairly important to 18.2%; very important to 15.3% and lastly, slightly important to 6.2%. Use of ICT to log queries is important to 33.5%, not important to 28.8% and very important to 19.5%; Fairly important to 10.6% and slightly important to 7.5%. Customer satisfaction evaluation through the use of ICT is not important to 31.2%, Important to 28.6% and fairly important to 19%; Very important to 15.6% and slightly important to 5.7%.

Both results show that ICT would be most useful for payments after trip completion. Least important use of ICT would be trip evaluation and evaluation of services after the trip. Customer satisfaction, especially in public transportation services is ignored. After journey completion, ICT would help both public transportation patrons and operators. Operators using feedback and reviews received from trip evaluation and queries from customers for service improvement. Customers would also benefit from improved services based on their feedback on ICT based platforms.

6   Summary and Conclusion

Current travel behaviour is not visibly influenced by ICT but there is scope for introduction of new ICT solutions to public and private transportation users in both cities. In both areas, the use of ICT for travel related reasons is seldom. However, there is interest in ICT use for ravel and transport reasons. The two cities have similar travel characteristics patterns. The only variation in available modes of transportation is the availability of buses in Bloemfontein and lack of buses in Gaborone. Access to activities is influenced by mobility provided by modes of transportation travellers have access to. ICT also has the potential to affect activities generating travel by way of travel substitution by opting for virtual access. Some work, educational and shopping trips can be replaced by tele-working, online schools and courses as well as online shopping, however there are jobs that cannot be replaced by ICT, therefore, travel substitution will not completely eliminate the need to travel for work. It is also essential to note the lack of flexibility in working times for full-time employees and students; which does not allow for remote virtual learning and working, therefore travel for these activities is less likely to be substituted by ICT. Compatibility of ICT, especially devices, and modes of transportation should also be considered.

Motorcycles and Non-Motorised modes, i.e. bicycles and walking have limited capacities for accommodating devices. Technological devices could also potentially be distractive to users. Use of devices such as mobile phones, music players and ear-phones have been linked to road accidents. Motorised vehicles have the potential to accommodate various devices, both personal to transportation users and built in to the vehicles. Transportation infrastructure; roads, traffic monitoring and control infrastructure, bus stops, taxi and bus rank infrastructure will also need to be upgraded to accommodate ICT components.

Overall, results indicated that regular travellers desire ICT for travel purposes. The need for information before beginning a trip is vital to trip making decisions and behaviour. Availability of information and access to ICT components alone does not change and impact travel behaviour. Cognitive and psychological factors in individuals play an important role in travel behaviour. Travellers attitudes and perceptions towards modes of transportation, their reasons for travel as well as ICTs would still impact their travel decision making (Ruiz *et al.* 2018).

The growth of technology based on-demand passenger transportation and ride-sharing platforms such as Uber, Lyft and Taxify has not reached the cities of Gaborone and Bloemfontein. These platforms growth over the last years have tremendously changed public transportation in most of the world's cities. However, their non-existence in Bloemfontein and Gaborone is an indication of lack of ICT integration in the transportation systems of these cities. Private vehicle users' travel behaviour have also been impacted by these platforms in ways such as opting for Uber instead of driving. At the final stages of a trip, ICT applications are available for tech-based public transportation service providers such as Uber and other ride-sharing platforms that allow clients to pay and rate their services using mobile applications. Evaluating the trip after completion, would be useful for public transportation users. This could help with improvement of services by the providers provided they consider feedback provided by users.

Accessibility of ICT as well as current uses shows possibilities of eased integration into transportation systems. In conclusion, even though there is limited ICT use and influence on travel, there is potential for ICT incorporation into the transportation system for travel purposes. The use of ICT would not necessarily eliminate the need to travel, but improve travel experiences.

References

de Abreu e Silva, J., de Oña, J., Gasparovic, S.: The relation between travel behaviour, ICT usage and social networks. The design of a web based survey. Transp. Res. Procedia. **24**, 515–522 (2017). https://doi.org/10.1016/J.TRPRO.2017.05.482

Ben-Elia, E., et al.: Activity fragmentation, ICT and travel: an exploratory path analysis of spatiotemporal interrelationships. Transp. Res. Part A Policy Pract. **68**, 56–74 (2014). https://doi.org/10.1016/j.tra.2014.03.016

Konrad, K., Wittowsky, D.: Research in transportation economics virtual mobility and travel behavior of young people – connections of two dimensions of mobility. Res. Transp. Econ. **68** (2017), 11–17 (2018). https://doi.org/10.1016/j.retrec.2017.11.002

Li, J., Lo, K.: Do socio-economic characteristics affect travel behavior? A comparative study of low-carbon and non-low-carbon shopping travel in Shenyang City, China (2018). https://doi.org/10.3390/ijerph15071346

Lila, P.C., Anjaneyulu, M.V.L.R.: Modeling the impact of ICT on the activity and travel behaviour of urban dwellers in indian context. Transp. Res. Procedia **17**(2014), 418–427 (2016). https://doi.org/10.1016/j.trpro.2016.11.083

Mokhtarian, P.L.: A typology of relationships between telecommunications and transportation. Transp. Res. Part A General. **24**(3), 231–242 (1990). https://doi.org/10.1016/0191-2607(90)90060-J

Nyblom, Å.: Making plans or "just thinking about the trip"? Understanding people' s travel planning in practice **35**, 30–39 (2014). https://doi.org/10.1016/j.jtrangeo.2014.01.003

Ruiz, T., et al.: Connotative meaning of travel modes and activity-travel behavior. Transp. Res. Procedia **33**, 379–385 (2018). https://doi.org/10.1016/j.trpro.2018.11.004

Snellen, D., De Hollander, G.: ICT'S change transport and mobility: mind the policy gap! Transp. Res. Procedia. **26**(2016), 3–12 (2017). https://doi.org/10.1016/j.trpro.2017.07.003

Statistics Botswana: Botswana in Figures 2016 (2016)

Varghese, V., Jana, A.: Impact of ICT on multitasking during travel and the value of travel time savings: empirical evidences from Mumbai, India. Travel. Behav. Soc. **12**(2017), 11–22 (2018). https://doi.org/10.1016/j.tbs.2018.03.003

Risk Analysis of Drivers' Distraction: Effect of Navigation Tools

Jacob Adedayo Adedeji[1]([✉]), Xoliswa E. Feikie[2],
and Mohamed M. H. Mostafa[3]

[1] Department of Civil Engineering, Central University of Technology,
Free State, Bloemfontein, South Africa
jadedeji@cut.ac.za
[2] Department of Civil Engineering and Geomatics,
Durban University of Technology, Durban, South Africa
[3] School of Engineering, Civil Engineering, University of KwaZulu-Natal,
Durban, South Africa

Abstract. Road users' characteristics are amongst the leading causes of traffic fatalities, leading to reduced levels of traffic safety. There are numerous characteristics of road users, yet, two of these characteristics standout, these include the visual acuity factor and the reaction process. There are various factors that contribute to the delay time and reaction process of drivers, and among these are the non-driving-related activities such as adjusting the stereo, environmental controls, conversing with passengers, using a cell phone, searching for street addresses, looking at/for a building…etc. However, there is not a lot of research on non-driving activities such as the use of navigation tools in the form of Global Positioning System (GPS) and navigation phone applications while driving. Through multiple data collection approaches, this study attempts to highlight the effect of navigation tools and risks involved when used while driving. Subsequently answering these questions on the drivers' reaction process; (i) Can navigation tools be classified as non-driving activities, (ii) Do navigation tools influence drivers' reaction and decision time and (iii) What are the possible remedies if it affects drivers' decision time. The findings of the study highlight the risk, the impact of navigation tools on drivers' behaviour, its influence on traffic crashes and propose the possible countermeasures to this effect.

Keywords: Traffic fatalities · Drivers' distractions · Behaviour · Traffic safety · Navigation tools

1 Driving and Advanced Technologies

Mobility is an essential part of modern society, leading to increased vehicles on the road. Nonetheless, global health estimates highlighted that road injury is part of the top ten global causes of deaths having killed 1.4 million people in 2016, about which 74% of whom were men and boys (WHO 2016). Some of the key contributing factors to this tragedy includes speed, drink driving, motorcycle helmets, seatbelts and child restraints and most especially distracted driving (WHO 2016; Stone et al. 2018). However, since the genesis of sustainable development goals, there has been a rapid development on

© Springer Nature Switzerland AG 2020
L. Mohammad and R. Abd El-Hakim (Eds.): GeoMEast 2019, SUCI, pp. 128–140, 2020.
https://doi.org/10.1007/978-3-030-34187-9_10

new technologies that have led to the manufacturing of mobile systems, which are sometimes adopted without considering all the implications and risks involved (Renaudin et al. 2016). Generally, driving in itself is a complex task, as it involves; navigation (planning and route following), guidance (following the road and maintaining a safe path in response to traffic conditions) and control (steering and speed control). Thus, drivers' distraction is not encouraged and all the factors that lead to the distraction of drivers should be prioritized when designing and adopting new systems and tools, but this is not always the case because these tools do not always follow a user-centred perspective before been commercialised (Ellison et al. 2015). Although human factors are taken into consideration during the design and introduction phase of these advanced technologies, are often not prioritized.

2 Navigation System and Ride-Sharing

In recent years, the Global Positioning System (GPS) and navigation phone applications are mostly used while driving, and these systems/tools have their limitations that can lead to the distraction of drivers, which consequently contribute to traffic fatalities that lead to reduced levels of safety on the road (Stone et al. 2018). Over the last few decades, the need for sustainability transportation has risen, and this gave birth of the era of ride-sharing and carpooling, this was in order to reduce traffic congestion and in turn the carbon emission footprint. Nevertheless, this birth was built on navigation tools and has further encouraged its use. Therefore, navigation tools such as Google Map, Waze, Navmii GPS South Africa, etc. are no longer a want but a need.

In South Africa, the use of Uber, Taxify and other ride-sharing app are on the increase, as this has not only solved the issue around mobility but have created employment for many. Consequently, the use of navigation tools has also increased and its advantages are numerous which includes; reduction of uncertainty and stress level, creation of shorter or alternative routes, provide information on traffic congestion and locate point of interest (Stone et al. 2018; Seshadri et al. 2009). Yet, the negative safety impact and behavioural changes caused by its usage cannot be ignored. Overall, the accuracy of a GPS is dependent on factors such as signal quality which is influenced by a number of visible satellites, surroundings and weather; while smartphones if used for the navigation applications, have limitations that include low accuracy; battery energy and limited processing power (Kanarachos et al. 2018). Thus, the goal of this current study is to analyse the effect of navigation tools on drivers' behaviour in terms of reaction and decision time form the perspective of the driver and passengers.

3 Methods

The approach used to collect data involved the use of questionnaires collected through an online platform QuestionPro. The expected feedback was from Africa, but the majority of the respondents were from specifically South Africa. The analysis of this paper consists of 203 respondents, the questionnaire used consists of two sections depending on whether a person is a driver or a passenger. The first section focuses on demographic characteristics of both drivers/passengers such as; age, gender, and educational level. The second section is different depending on whether a respondent is a driver or a passenger. Passengers' questions composed of multiple choice questions focusing on the types of commercial transportation used and the first-hand challenges experienced while using their respective commercial transportations. The drivers' questions consisted of multiple choice questions focused on the type of driver's license they obtained, the purpose of their driving and the duration of years they have been driving including the number of traffic fines they have received in the past year/s. The other part of the questionnaire focused on the types of drivers they are in terms of paying attention to certain details while driving such as safety belts, safety driving distances, and maintaining speed limits along driving lanes. The last part focused on the use of the navigation tools, their expertise in using the navigation tool on the Likert-type scale, as well as factors that contribute to their distraction while driving.

The use of chi-square statistical tool was employed to test for the hypothesis relating to the road user's perception of navigation tools. Thus, the null and alternative hypotheses for the testing were:

H$_o$: The road user's characteristics and perception of navigation tools as driver's distraction are independent of each other.

H$_i$: Null hypothesis is not true

The expected cell frequencies were compared with the observed cell frequencies using the test chi-square, as estimated.

$$X^2 = \sum \frac{(O_i - E_j)^2}{E_{ij}}$$

where:

X^2 = chi-square

O_{ij} = observed frequency of the cell in the ith row and jth column

E_{ij} = expected frequency of the cell in the ith row and jth column

The calculated chi-square result was compared with the critical chi-square value (using the table) with $(r-1) \times (c-1)$ degree of freedom to make a decision regarding the acceptance or rejection of the null hypothesis, Kothari (2004).

3.1   Decision Rule

If $X^2_{tab} > X^2_{cal}$, accept H$_o$, otherwise reject.

4   Results/Findings

4.1   Demographic Information of Respondents

A total of 203 respondents responded to the questionnaire, 52% were passengers and the remaining were drivers, and from the total respondents, 58% were male (Table 1). Most (65%) of the respondents are between the ages of 18 to 30 and only 9% are above

40 years old. Educational level among the respondents' ranges from undergraduate, postgraduate and grade 1–12 at 52%, 30% and 18% respectively, this will have resulted that the survey was distributed to universities students. Although, 60% of the respondents own a driver's license however, on a regular basis 47% are drivers which imply are in custody of cars to drive.

Table 1. Road users demographic data

Demographic characteristics	Class	Percentage
Road users	Passengers	52
	Drivers	48
Age	From 18 to 25	44.0%
	From 26 to 30	31.0%
	From 31 to 39	17.0%
	From 40 to 49	6.0%
	Above 50	2.0%
Gender	Female	42.0%
	Male	58.0%
Education	Grade 1–6	5.0%
	Grade 7–12	13.0%
	Undergraduate	52.0%
	Postgraduate	30.0%
Owning driver license	Yes	60.0%
	No	40.0%
On a regular basis, are you	Driver	47.0%
	Passenger	53.0%

4.2 Drivers Data and Navigation Tools

Out of the 47% that are drivers on regular basis, 89% are in possession of South Africa driver's license and with the majority (41%) of the drivers are licensed for a light motor vehicle (license type B) followed by the licence type C1 (heavy vehicle up to 16 000 kg GVM) (Table 2). Additionally, the majority of the respondents drive for private use and only 8% drive for commercial. A minority (20%) of the drivers interviewed have been driving for a while (11 to 40 years), while the majority have been driving for 4 to 10 years, and novice drivers about 30%. The driving experience is set to influence their perception of the navigation tools, as it is expected to first master the act of driving and seconding understanding the use of navigation tools. Over the last three years, about 29% have been involved in a car accident, this can be some worth justified with the percentage of novice drivers. Also, only 41% have never received fines for violation of traffic rules within the past year.

Table 2. Drivers data and navigation tools

Drivers data	Class	Percentage
Driver license issuer	South Africa	89.0%
	Other countries	11.0%
License type	A1	2.0%
	A	2.0%
	B	41.0%
	EB	8.0%
	C1	39.0%
	C	8.0%
	Others	2.0%
Driving purpose	Commercial	8.0%
	Private	92.0%
Type of commercial	Ride-Hailing	17.0%
	Meter-Taxi	33.0%
	Other	50.0%
Years of driving license	0 to 3	30.0%
	4 to 10	50.0%
	11 to 25	15.0%
	26 to 40	5.0%
Involved in a car accident within the last 3 years	No	71.0%
	Yes	29.0%
Driving fines	0	41.0%
	1 to 3	47.0%
	4 to 6	6.0%
	7 to 10	3.0%
	Over 10	3.0%
Distracted while driving	Always	4
	Often	13
	Sometimes	41
	Rarely	34
	Never	8
Using a mobile phone while driving	Always	12
	Often	17
	Sometimes	44
	Rarely	17
	Never	11
How often do you use a navigation tool	Always	12
	Often	23
	Sometimes	37
	Rarely	22
	Never	6

(*continued*)

Table 2. (*continued*)

Drivers data	Class	Percentage
Level of expertise in using the navigation tools	Excellent	20
	Good	56
	Fair	18
	Poor	5
	Very poor	1

Furthermore, from Table 2, only 8% of the interviewed drivers are 'never' distracted while driving, although, over 70% of the drivers sometimes use their mobile phone while driving. Consequently, never distracted could be justified based on the 20% experienced drivers. Most of the drivers interviewed at 37% 'sometimes' use the navigation tools and 28% rarely use it. Yet, 76% claimed to have a good to excellent knowledge in using the navigation tools, this correlates to the fact that only 30% of the drivers are novice and majority of the respondents are educated.

4.3 Passengers Data and Navigation Tools

With regards to the passengers (Table 3), the study indicated that only 35% of the passengers have driver's license but are a passenger regularly, this implies that they do not have a car at their possession. Thus, the majority of the passengers use public transportation, with 26% using minibuses, 27% meter taxies and 43% ride-hailing taxies. Additionally, 46% of the respondents have never or rarely use the ride-hailing App, and the majority use the App for unfamiliar destinations. Overall, the passenger prefers the drivers to use navigation tools than being guided by the driver on an unknown route; this implies that the passengers prefer the judgement of the navigation tools over the drivers' judgement of direction. Although, 44% responded that any option is fine, but this has resulted in a lot of conflicts between drivers and passengers as the passenger tend to blame the driver for taking a longer route to increase the fare.

Table 3. Passengers data and navigation tools

Passengers data	Class	Percentage
Owning driver license	No	69
	Yes	35
How often do you use commercial transport	Always	20
	Often	31
	Sometimes	29
	Rarely	11
	Never	9

(*continued*)

Table 3. (*continued*)

Passengers data	Class	Percentage
Which commercial transportation system do you use the most	Uber	18
	Bolt (taxify)	25
	Meter Taxi	27
	Minibus	26
	None	5
How often do you use the Ride-hailing App	Always	4
	Often	12
	Sometimes	38
	Rarely	20
	Never	26
What activities do you use the Ride-hailing App for?	School	25
	Work	13
	Leisure	25
	Unfamiliar destination	38
What do you expect of a driver on unknown route?	I prefer to guide	13
	I prefer Driver to use navigation tools	43
	Any is fine	44

4.4   Road Users Perception of Navigation Tools to Danger

Figures 1 and 2 show the perception of road users in terms of impact of navigation tools to drivers distraction and drivers error or mistake. Results show a slight balance in the perception of road users towards the contribution of navigation tools to the driver's distraction. Nevertheless, the driver's perception when compared to that of the passenger differs, as driver tend to agree (18% Always and 24% often get distracted) that navigation tools increase the probability of been distracted and the majority of the passenger's opinion seem indifferent. This can be justified base on the fact that passengers might not have discovered the risk posed to drivers when using a navigation tools. The risk further increase with the failure associated to navigation tools such as signal quality, low accuracy; battery energy, limited processing power and other inferring pop ups (such as message notifications and phone calls) when using a phone App navigation tools. On the other hand, the majority of the road users disagree that navigation tools contribute to drivers' error or mistake (Fig. 2). This may be justified on the fact that driver's error or mistake entails a great deal of other factors which can be fatigue, negligence, lack of understanding of traffic signs and regulation and even vehicle faults.

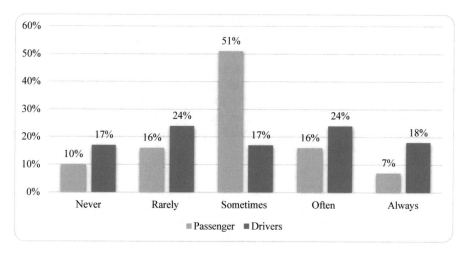

Fig. 1. Road users perception of navigation tools contribution to driver's distraction

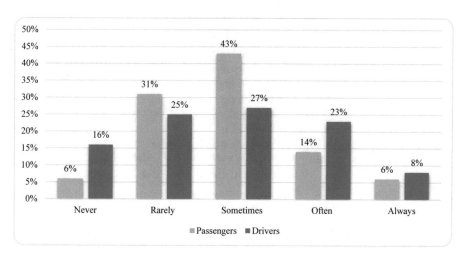

Fig. 2. Road users perception of navigation tools contribution to driver's error or mistake

Furthermore, the perception of the road users on the other causes of distraction was evaluated, the results show that there is similarity in the perception for the drivers and the passengers of these factors (Fig. 3). Firstly, text messaging (Texting and Driving) at 18% and 17% by drivers and passengers respectively is seen as the highest risk of distraction to drivers, this result agrees with the fact that 25% of accidents relate to the use of cell phone while driving (Tedesco 2014). Nevertheless, text messaging is one of the means of communicating with the driver on the ride-hailing App, which may further increase the probability of drivers being distracted while driving. Secondly, conversing with the passenger(s) while driving is also seen as a form of distraction, with the driver's perception at 11% and passenger's at 13%, although, previous studies disagree with that, stating that it improves driving performance (Drews et al. 2008; Maciej et al. 2011). Nevertheless, the conversation pattern of the drivers is in question needs to be examined. Thirdly, eating or drinking while driving is considered more as a form of distraction by the drivers and this might be from the perception of being an unprofessional act especially when the passenger(s) are on-board. Furthermore, the passengers at 19% and 16% against the drivers 11% and 12% perceived driver talking over the phone and dialing/hanging up on the phone respectively as a form of distraction, as the driver on a phone conversation pays less attention to the surrounding traffic which can result in traffic fatality.

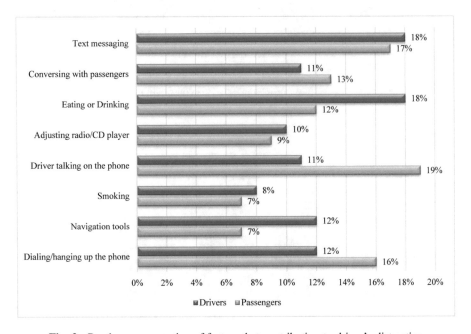

Fig. 3. Road users perception of factors that contributing to driver's distraction

4.5 Cross Analysis of Road Users and Driver Distracted Driving

Evaluation of the three questions was conducted to check the relationship between the road user's perception of navigation tools to driver's distraction and road user's characteristics with regards to the different types of road users, gender, age, educational background and drivers' years of driving experience.

4.5.1 Cross-Classification Analysis of Road Users and Perception of Effect of Navigation Tools

In the cross-classification analysis of road users and the perception of the effect of navigation tools, the results are presented in Table 4. Considering 5% level of significance, the chi-square value is 9.49. However, the calculated chi-square values for the perception of navigation tools to driver's distraction were larger than the critical

Table 4. Cross analysis with road users

Road users	Perception of navigation tools to the driver's distraction				
	Never	Rarely	Sometimes	Often	Always
Passengers	10	16	51	16	7
Drivers	17	24	17	24	18
Chi-square test statistic	26.86				
H$_o$ rejected?	Yes				

value, thus, indicating that there are some reasons to believe that the variables are dependent. Additionally, since the $X^2_{cal} > X^2_{tab}$, reject the H$_o$, meaning the different type of road users and their perception of navigation tools to drivers distraction are dependent on each other.

4.5.2 Cross-Classification Analysis of Genders and Perception of Effect of Navigation Tools

Table 5 shows that statistical result of the association between gender and their perception of the effect of navigation tools. Considering 5% level of significance, the chi-square value is 9.49 but the calculated chi-square at 7.03. this implies that there is no relationship between gender and their perception of navigation tools to drivers distraction.

Table 5. Cross analysis with genders

Genders	Perception of navigation tools to the driver's distraction				
	Never	Rarely	Sometimes	Often	Always
Female	6	13	29	20	13
Male	20	27	42	22	11
Chi-square test statistic	7.03				
H$_o$ rejected?	No				

4.5.3 Cross-Classification Analysis of Age and Perception of Effect of Navigation Tools

Table 6 presents the cross-classification analysis of age and the perception of the effect of navigation tools. The result shows that the chi-square value is 5.76 and considering a 5% level of significance the chi-square value is 26.30, the calculated chi-square is less than the critical value, thus, indicating that the variables are independent. This implies that road users age and their perception of navigation tools to drivers distraction are independent of each other.

Table 6. Cross analysis with age

Age	Perception of navigation tools to the driver's distraction				
	Never	Rarely	Sometimes	Often	Always
From 18 to 25	5	9.5	17.5	6.5	3.5
From 26 to 30	5	5	9.5	7	6.5
From 31 to 39	1	3	4.5	4.5	1.5
From 40 to 49	1.5	1.5	2.5	1.5	0.5
Above 50	0.5	0.5	1	0.5	0.5
Chi-square test statistic	5.76				
H_o rejected?	No				

4.5.4 Cross-Classification Analysis of Educational Background and Perception of Effect of Navigation Tools

In the cross-classification analysis of the educational background and the perception of the effect of navigation tools, the results are presented in Table 7. Considering 5% level of significance, the chi-square value is 21.03. However, the calculated chi-square values for the perception of navigation tools to driver's distraction were larger than the critical value, thus, indicating that there are some reasons to believe that the variables are dependent. Additionally, since the $X^2_{cal} > X^2_{tab}$, reject the H_o, meaning educational background and their perception of navigation tools to drivers distraction are dependent on each other.

Table 7. Cross analysis with educational background

Educational background	Perception of navigation tools to the driver's distraction				
	Never	Rarely	Sometimes	Often	Always
Grade 1–6	0	0	9	3	3
Grade 7–12	9	2	8	7	2
Undergraduate	16	18	38	18	10
Postgraduate	4	19	20	10	7
Chi-square test statistic	26.37				
H_o rejected?	Yes				

4.5.5 Cross-Classification Analysis of Years of Driving License of Drivers and Perception of Effect of Navigation Tools

Table 8 shows that statistical result of the association between drivers' years of driving experience and their perception of the effect of navigation tools. Considering 5% level of significance, the chi-square value is 21.03 but the calculated chi-square at 10.6. Therefore, it implies that there is no relationship between drivers' years of driving experience and their perception of navigation tools to distraction.

Table 8. Cross analysis with years of driving license of drivers

Years of driving license of drivers	Perception of navigation tools to the driver's distraction				
	Never	Rarely	Sometimes	Often	Always
0 to 3	6	9	3	7	7
4 to 10	5	14	8	13	9
11 to 25	3	4	5	3	0
26 to 40	1	2	0	0	1
Chi-square test statistic	10.6				
H$_o$ rejected?	No				

Overall, road users in terms of driver and passenger, and educational background have a significant influence on the perception of navigation tools to the driver's distraction. However, gender, age of the road users and drivers' years of driving experience do not influence their perception of navigation tools. It can be deduced that drivers and passengers reasoning differs on the issue related to navigation tools. Also, the inbuilt text messaging and phone call system of the ride-hailing App should be regulated in such a way that it can be disable when driver is in motion. This will further reduce the probability of drivers been distracted.

In addition, educational background enhances road users level of expertise in using that navigation tools. However, this expatiate the acceptance and use of navigation tools and other related tools in developed cities with high level of literacy. Nevertheless, the complications around the interaction mechanism of any driving related tools should be minimized. On the other hand, implementation of different languages and standardizing symbols use in navigation tools should be encourage.

5   Conclusion

This study focus on the analysis of contribution of navigation tools to drivers distraction, thus answering these research questions. *Can navigation tools be classified as non-driving activities;* research show that there are various non-driving related activities but as a result of the necessity for navigation tools, it is gradually becoming more driving-related activities. Coupled with the response of the passengers preference of drivers to use navigation tools at unknown route. Also, navigation is one of the critical task of a driver alongside with guiding and controlling, thus, navigation tools should enhance drivers driving task. This was alluded in the response of the road users and it should not be seen as a non-driving related activities.

Furthermore, *do navigation tools influence drivers' reaction and decision time;* results show that 12% of the drivers and only 8% of the passengers agree that navigation tools influence drivers distraction (which influence drivers' reaction and decision time). Nevertheless, there are other factors which are high-risk contributors to drivers distraction such as text messaging, eating and drinking, drivers talking on the phone. Finally, *what are the possible remedies if it affects drivers' decision time;* although, research shows that there are other factors that are riskier when compared with navigation tools, yet, navigation tools most time aided the use their usage. For instance, using a cell phone-based navigation tools will allow drivers to notice other phone related activities such as text message, incoming calls ...etc. which contributes to drivers distraction. Thus, remedies such as phone disabling text messaging and phone call system when drivers are on motion and encouraging the inbuilt ride-hailing system for vehicles. Overall, navigation tools might not have contributed greatly or directly to drivers distraction but the use of phone-based navigation tools have contributed to drivers distraction.

References

Drews, F.A., Pasupathi, M., Strayer, D.L.: Passenger and cell phone conversations in simulated driving. J. Exp. Psychol. Appl. **14**(4), 392 (2008)

Ellison, A.B., Greaves, S.P., Bliemer, M.C.: Driver behaviour profiles for road safety analysis. Accid. Anal. Prev. **76**, 118–132 (2015)

Kanarachos, S., Christopoulos, S.R.G., Chroneos, A.: Smartphones as an integrated platform for monitoring driver behaviour: the role of sensor fusion and connectivity. Transp. Res. Part C Emerg. Technol. **95**, 867–882 (2018)

Kothari, C.R.: Research Methodology-Methods and Techniques, 2nd edn. New Age International Publishers, New Delhi (2004)

Maciej, J., Nitsch, M., Vollrath, M.: Conversing while driving: the importance of visual information for conversation modulation. Transp. Res. Part F Traffic Psychol. Behav. **14**(6), 512–524 (2011). https://doi.org/10.1016/j.trf.2011.05.001

Renaudin, V., Dommes, A., Guilbot, M.: Engineering, human, and legal challenges of navigation systems for personal mobility. IEEE Trans. Intell. Transp. Syst. **18**(1), 177–191 (2016)

Seshadri, N., Karaoguz, J., Broadcom Corp.: Sharing of GPS information between mobile devices. U.S. Patent Application 12/026,582 (2009)

Stone, M., Knapper, J., Evans, G., Aravopoulou, E.: Information management in the smart city. Bottom Line **31**(3/4), 234–249 (2018)

Tedesco, C.: Cell Phone Use While Driving. Policy Memorandum Institute for Public Policy Studies, University of Denver (2014)

World Health Organization: World health statistics 2016: monitoring health for the SDGs sustainable development goals. World Health Organization (2016)

Author Index

© Springer Nature Switzerland AG 2020
L. Mohammad and R. Abd El-Hakim (Eds.): GeoMEast 2019, SUCI, p. 141, 2020.
https://doi.org/10.1007/978-3-030-34187-9

Printed in the United States
By Bookmasters